Strong Stable Markov Chains

STRONG STABLE MARKOV CHAINS

N.V. Kartashov

///VSP///

≪TBiMC≫

Utrecht, The Netherlands 1996 Kiev, Ukraine

VSP BV
P.O. Box 346
3700 AH Zeist
The Netherlands

First published in 1996

ISBN 90-6764-205-3

Printed in The Netherlands by Koninklijke Wöhrmann BV, Zutphen.

CONTENTS

PREFACE

This monograph presents a new approach to the investigation of ergodicity and stability problems for homogeneous Markov chains with a discrete-time and with values in a measurable space. The main purpose of this book is to highlight various methods for the explicit evaluation of estimates for convergence rates in ergodic theorems and in stability theorems for wide classes of chains. These methods are based on the classical perturbation theory of linear operators in Banach spaces and give us new results even for finite chains.

Let $X = (X_t, t \geq 0)$ be a homogeneous discrete-time Markov chain with values in a measurable state space $(E, \mathcal{E})$. From the analytical point of view, any chain can be considered as its transition kernel P on $(E, \mathcal{E})$. This kernel defines one-step ahead probabilities. A linear operator defined in the usual way on the Banach space $m\mathcal{E}$ of finite signed measures on $\mathcal{E}$ corresponds to the kernel P.

Using this relationship, the theory of quasicompact chains was developed over the last four decades by means of the linear operator theory, [Doeblin (1940), Yosida and Kakutani (1941), Bogolyubov and Krein (1946), Krein and Rootman (1948), Hille and Phillips (1957); see also Neveu (1964), Chapter 5; Revuz (1975), Chapter 6]. Quasicompact chains are often also called uniformly ergodic, uniformly recurrent, strong positive recurrent, or uniformly mixing chains. Uniformly ergodic chains possess some important properties and are widely used in applications: see Doob (1953), Nagaev (1961), Davydov (1973), Brunel and Revuz (1974), Zubkov (1979), Kalashnikov (1981), Anisimov (1988), Korolyuk and Turbin (1976), (1978), (1982). In particular, the uniform recurrence of a chain can be easily checked by a simple criterion for the quasicompactness, such as the uniform mixing condition or the Doeblin condition, that is expressed in terms of its transition kernel P. This fact makes simpler the checking of corresponding conditions and also allows us to obtain bounds for the convergence in the form of explicit inequalities. Such bounds are uniform on wide classes of Markov chains — because an "ergodicity index" of a quasicompact chain can be effectively expressed in terms of its kernel P.

Often we do not know precisely the transition probabilities and only estimates for some moment functionals of a chain are available. For this reason we must use only uniform limit theorems that are valid in certain neighborhoods of the transition operator. Therefore, uniform bounds obtained below are of a certain interest, and especially in statistics.

Ergodic properties of the quasicompactness follow from the simple structure of the spectrum of the linear operator P. It is clear that this spectrum depends not only on the analytical definition of the kernel P but also on properties of the Banach space $m\mathcal{E}$. We show that a restriction of the domain of the operator P from $m\mathcal{E}$ to some P-invariant subspace $\mathfrak{M} \subset m\mathcal{E}$ with a norm $\|\cdot\|$ can essentially improve the spectral characteristics of the operator P. Often this approach allows us to obtain ergodic properties of a chain that is not quasicompact from ones of a quasicompact chain. The investigation of the quasicompactness in this extended sense is our first goal.

Another purpose of the monograph is to introduce and investigate the stability property of a Markov chain. Each chain can be considered as a map of its transition operator to a set of finite-time distributions, ergodic distributions and other characteristics, that are calculated by one-step transition probabilities. The chain with the operator P is said to be strong stable if this map is continuous at the point P under suitable choice of metric. In other words, small perturbations of a transition operator of a strong stable Markov chain induce only small changes of finite-time and ergodic distributions. This notion is closely related to the stability theory of stochastic systems developed over the last decades by Zolotarev (1976), Borovkov (1977), and Kalashnikov (1983).

The first part of the monograph develops the theory of uniform ergodic chains with respect to a given norm. This theory is based on the approach described above and contains all basic results of the theory of quasicompact chains including period, cyclic subclasses and Doeblin's criteria.

Under certain assumptions on the norm, the equivalence between uniform ergodicity and strong stability is established. We give explicit uniform bounds for the convergence, introduce periodicity classes, and investigate an asymptotics of transition probabilities. Asymptotic expansions with explicitly evaluated correction terms in ergodicity and stability theorems are also obtained including uniform estimates of a remainder.

In the second part, we remove the condition on the uniform ergodicity. Actually, the unique condition on the initial non-perturbed chain X is the existence of a unique finite invariant measure or the finiteness of any invariant measure. Instead of the ergodicity condition, we assume that the perturbation D of the transition operator P must be subordinated to the generalized potential of the chain, that is, to an unbounded operator with its own domain. We show that this condition, without additional assumptions, is necessary and sufficient for stability of the ergodic distribution. Particularly, if X is uniformly ergodic then this condition can be reduced to the boundedness of the perturbation D in a chosen space $\mathfrak{M}$ of measures. The main part of the induced "stability" norm of a perturbation is expressed in terms of a nonnegative operator generated by means of hitting times. We also obtain explicit estimates for the stability as well as asymptotic expansions in stability theorems.

Based on the analytical methods proposed by Korolyuk and Turbin (1976-1982), we consider the problem of the states consolidation of a Markov chain as a special case of the stability problem. Explicit inequalities, asymptotic expansions, and uniform consolidation theorems are obtained. Exponential asymptotics of "rare" Markov moments on chains are also investigated and corresponding explicit bounds are given.

The main results of the first two chapters are illustrated for the problem of the stability and the ergodicity of a waiting times sequence in a one-channel queuing system under small perturbations of interarrival and service times of a general form.

For instance, the linear correction term of a distribution of the steady-state waiting time under such perturbations is evaluated.

Some new inequalities are given in Chapter 9 for the distance in the uniform metric of the distribution function of a sum of a random geometric number of independent identically distributed nonnegative random variables and the corresponding exponential function. The right-hand sides of those estimates have the first order of smallness with respect to the small parameter of a geometric distribution and contain also the second moment of the terms. The question of the best possible constants is considered.

Estimates of the convergence rate in the classical Renyi theorem are also given.

CHAPTER I

UNIFORMLY ERGODIC AND STRONG STABLE MARKOV CHAINS

We introduce here basic assumptions and list properties of the domain of definition $\mathfrak{M}$ of a linear operator generated by a transition kernel P. Further we give a characterization of the uniform ergodicity of a chain in terms of the generalized potential and the resolvent of the kernel P. Then we establish the equivalence with respect to the same norm between the uniform ergodicity and the strong stability of the invariant distribution of a chain.

1.1. Norms in spaces of measures, functions, and kernels

Let $(E, \mathcal{E})$ be a measurable space. Denote by $m\mathcal{E}$, $f\mathcal{E}$, and $b\mathcal{E}$ the spaces of finite signed measures on $\mathcal{E}$, measurable functions on $\mathcal{E}$, and measurable bounded functions on $\mathcal{E}$, respectively. Let $m\mathcal{E}^+$, $f\mathcal{E}^+$, and $b\mathcal{E}^+$ be the cones of nonnegative elements in these spaces.

For all transition kernels $Q(x, A)$, $x \in E$, $A \in \mathcal{E}$, and for all measurable functions $f(x)$, $x \in E$, we define linear mappings $Q: f\mathcal{E} \to f\mathcal{E}$, $Q: m\mathcal{E} \to m\mathcal{E}$, $f: m\mathcal{E} \to \mathbb{R}$ as follows

$$\begin{aligned} Qg(x) &= \int Q(x, dy)g(y), \\ \mu Q(A) &= \int \mu(dx)Q(x, A), \\ \mu f &= \int \mu(dx)f(x) \end{aligned} \tag{1.1}$$

provided that these integrals are well defined.

Here and in the sequel the integral, the sum, and the supremum signs without specifications on a domain are considered on a whole domain of definition of corresponding variables.

Let the symbol $f \circ \mu$ stand for the direct product $f(x)\mu(A)$, $x \in E$, $A \in \mathcal{E}$. As usual, $|\mu|$ means the full variation of a measure μ.

Assume that a Banach space $\mathfrak{M}$ is given in $m\mathcal{E}$ with a norm $\|\cdot\|$ such that

(M1) $|\mu(E)| \le k\|\mu\|$ for all $\mu \in \mathfrak{M}$ and some k.

For the subspace $\mathfrak{M}$, we introduce the dual space $\mathfrak{N}$ of functions on $f\mathcal{E}$ with the finite norm

$$\|f\| = \sup\{|\mu f|, \|\mu\| \le 1\}. \tag{1.2}$$

Such functions are supposed to coincide up to the equivalence: $f = g$ if and only if $\mu f = \mu g$ for all $\mu \in \mathfrak{M}$. We also introduce the space $\mathfrak{B}$ of transition kernels Q such that $\mathfrak{M}Q \subset \mathfrak{M}$ and the following operator norm is finite

$$\|Q\| = \sup\{\|\mu Q\|, \|\mu\| \le 1\}. \tag{1.3}$$

It follows from condition (M1) that the function $\mathbb{I} \in b\mathcal{E}$, that is equal to the unity identically, belongs to $\mathfrak{N}$ and $\|\mathbb{I}\| \le K$. Note that the operator norms introduced in

(1.3) for nonequivalent initial norms on $m\mathcal{E}$ are in the general case not comparable, even in the case where one of them is subordinated with respect to another.

In what follows we treat a norm of a function or of an operator as one induced by (1.2) or (1.3). Transition kernels and corresponding linear operators are denoted by the same symbols.

For two kernels P and Q, we define their product PQ as the kernel

$$PQ(x,A) = \int P(x,dy)Q(y,A).$$

We denote by P^t the t-times product of P by itself. By the definition we put $P^0 = I$ where I is the unit operator in $\mathfrak{B}$, that is, a kernel of the form $\mathbb{I}_{x\in A}$.

It follows from definitions (1.2) and (1.3) that [see Dunford and Schvartz (1953) or Rudin (1973)]

$$\|\mu Q\| \le \|\mu \cdot \|Q\|, \qquad \|Qf\| \le \|Q\| \cdot \|f\|,$$
$$|\mu f| \le \|\mu\| \cdot \|f\|, \qquad \|Q\| = \sup\{\|Qf\|, \|f\| \le 1\}.$$

We assume throughout that condition (M1) is valid as well as usual properties of a norm in a Banach space are satisfied. In addition, we use the following consistency condition on the norm and the order structure in $m\mathcal{E}$

(M2) $\|\mu_1\| \le \|\mu_1 + \mu_2\|$ for $\mu_i \in \mathfrak{M}^+$;

(M3) $\mathfrak{M}$ is a substructure in $m\mathcal{E}$, that is, $\|\mu_1\|, |\mu| \in \mathfrak{M}$ for $\mu \in \mathfrak{M}$ and $\|\mu_1\| \le \|\mu_1 - \mu_2\|$ for $\mu_i \in \mathfrak{M}^+$, $\mu_1 \perp \mu_2$.

Here and in the sequel $\mathfrak{M}^+$ and $\mathfrak{N}^+$ are the cones of nonnegative elements in the spaces $\mathfrak{M}$ and $\mathfrak{N}$ generating the orders in $\mathfrak{M}$, $\mathfrak{N}$, and $\mathcal{B}$. By definition, $\mu \le \nu$ if and only if $\nu - \mu \in \mathfrak{M}^+$, $f \le g$ if and only if $g - f \in \mathfrak{N}^+$, and $P \le Q$ if and only if $Q - P \in \mathcal{B}^+ = \{T \in \mathcal{B}: \mathfrak{M}^+ T \subset \mathfrak{M}^+\}$.

We also use the condition

(M3′) $\|\mu_1 - \mu_2\| = \|\mu_1 + \mu_2\|$ for $\mu_i \in \mathfrak{M}^+$, $\mu_1 \perp \mu_2$.

The last condition means that the measures μ and $|\mu|$ have the same norm in $\mathfrak{M}$. Analogously, we use the condition for the dual cone

(M2′) $\|f_1\| \le \|f_1 + f_2\|$ for $f_i \in \mathfrak{N}^+$.

This condition evidently follows from (M1)–(M3).

Under (M2) and (M3), condition (M1) yields

(M1′) $|\mu|(E) \le 2k\|\mu\|$ for all $\mu \in \mathfrak{M}$ and some k.

Note that the constant k may be used here instead of $2k$ if (M3′) is valid.

Conditions (M1)–(M3) and (M1′)–(M3′) are satisfied, in particular, for the following spaces $(\mathfrak{M}, \|\cdot\|)$. Let a measurable function v on E be such that $\inf v > 0$. Define, for all $\mu \in m\mathcal{E}$, a weighted variation norm

$$\|\mu\|_v = \int v(x)\, |\mu|(dx), \tag{1.4}$$

and set $\mathfrak{M}_v = \{\mu \in m\mathcal{E}: \|\mu\|_v < \infty\}$. Corresponding norms in $\mathfrak{N}_v$ and $\mathcal{B}_v$ are of the form [see Goldenstein, Gohberg, and Markus (1957) or Krein at al. (1972)]

$$\|f\|_v = \sup\left\{\frac{|f(x)|}{v(x)}, x \in E\right\},$$
$$\|Q\|_v = \sup\left\{\int |Q|(x,dy)\frac{v(y)}{v(x)}, x \in E\right\}. \tag{1.5}$$

Let $q \geq 1$ and let φ be a probability measure on $\mathcal{E}$. Put for $\mu \ll \varphi$

$$\|\mu\|_{q\varphi} = \left(\int \left|\frac{d\mu}{d\varphi}\right|^q d\varphi\right)^{1/q} \tag{1.6}$$

and $\mathfrak{M}_{q\varphi} = \{\mu \in m\mathcal{E}: \mu \ll \varphi, \|\mu\|_{q\varphi} < \infty\}$. Then the corresponding space $\mathfrak{N}_{q\varphi}$ coincides with $L^p(\varphi)$ where $p^{-1} + q^{-1} = 1$.

Due to Krein et al. (1972), chapter 8, from (M2) follows that the cone $\mathfrak{M}^+$ is normal and from (M3) follows that this cone is reproducing in $\mathfrak{M}$.

1.2. Uniformly ergodic chains

Let $X = (X_t)$ be a homogeneous Markov chain with values in a measurable space $(E, \mathcal{E})$ and with a discrete time $t = 0, 1, \ldots$ that is given by its transition kernel $P(x, A)$, $x \in E$, $A \in \mathcal{E}$.

Denote by $P^t(x, A)$ the transition probabilities over t steps. The kernel P^t is a t-fold power of P. Define also Cesaro averages $P^{(t)} = \sum_{s<t} P^s/t$, where $P^0 = P^{(0)} = I$ and I is the unit operator in $\mathfrak{M}$.

Introduce the following condition

(P) $\mathfrak{M}P \subset \mathfrak{M}$ and $\|P\| < \infty$.

Under condition (P), $P^{(t)}$ are linear bounded operators on $\mathfrak{M}$ and, therefore, the following definition is correct.

DEFINITION 1.1. A chain X is called uniformly ergodic with respect to a given norm $\|\cdot\|$ if there exists a stochastic kernel Π such that $P^{(t)} \to \Pi$ as $t \to \infty$ in the induced operator norm (1.3).

REMARK 1.1. A uniformly ergodic with respect to a norm chain may not have this property with respect to another one, even in the case where one of these measures is subordinated to the another.

DEFINITION 1.2. A stochastic kernel Π is called a stationary projector of a kernel P on $(E, \mathcal{E})$ if

$$\Pi^2 = \Pi = P\Pi = \Pi P \tag{1.7}$$

and $\mu = \mu\Pi$ provided that $\mu = \mu P$, $\mu \in \mathfrak{M}$.

THEOREM 1.1. *Let a chain X be uniformly ergodic with respect to a norm $\|\cdot\|$. Then the limit kernel Π in Definition 1.1 is the stationary projector of the kernel P.*

PROOF. It is easily seen that $\left(1 + t^{-1}\right) P^{(t+1)} - t^{-1}I = PP^{(t)} = P^{(t)}P$. Passing to the limit as $t \to \infty$ we conclude that $\Pi = P\Pi = \Pi P$. Therefore $\Pi^2 = \lim \Pi P^{(t)} =$

$\lim \Pi = \Pi$. If $\mu = \mu P$ then $\mu = \mu P^{(t)} \to \mu\Pi$ as $t \to \infty$ which is what had to be proved. □

REMARK 1.2. If a kernel P has a unique invariant measure π then P has a unique stationary projector Π of the form $\Pi = \mathbb{I} \circ \pi$, that is, $\Pi(x, A) = \pi(A)$ for all $x \in E$, $A \in \mathcal{E}$. Indeed, since $\mu\Pi = \mu\Pi P$, $\mu\Pi \in \{a\pi\}$ for all μ, that is, $\Pi = f \circ \pi$ for some $f \in \mathfrak{N}$. Since $\Pi\mathbb{I} = \mathbb{I}$ by the definition, $f = \mathbb{I}$ and $\Pi = \mathbb{I} \circ \pi$. If in addition $\pi \in \mathfrak{M}$ then $\Pi \in \mathcal{B}$.

REMARK 1.3. Following Korolyuk and Turbin (1978) we assume that in the space E there exists a partition $E = \bigcup_{u \in \mathfrak{U}} E_u$, $E_u \in \mathcal{E}$, such that, for some $\mathfrak{U}_1 \subset \mathfrak{U}$, the sets E_{u_1} are minimal ergodic classes for the chain X and the class $E_0 = \bigcup_{u \in \mathfrak{U} \setminus \mathfrak{U}_1} E_u$ contains zero states, that is,

$$\lim_{t \to \infty} \sup_{x \in E_0} \mathsf{P}\left(X_t \in E \setminus E_0 / X_0 = x\right) = 1$$

and, for all $u \in \mathfrak{U}_1$, $v \in \mathfrak{U} \setminus \mathfrak{U}_1$, the function

$$q(x, u) = \mathsf{P}\Big(\bigcup_{n \geq 1} \bigcap_{t \geq n} \{X_t \in E_u\} / X_0 = x\Big)$$

does not depend on x for $x \in E_v$.

Then, as is shown in Korolyuk and Turbin (1978), Chapter 6, the chain X has a stationary projector Π of the form:

$$\Pi(x, A) = \begin{cases} \pi_u(A), & \text{for } x \in E_u,\ u \in \mathfrak{U}_1, \\ \sum_{v \in \mathfrak{U}_1} q(x, v)\pi_v(A), & \text{for } x \in E_u,\ u \in \mathfrak{U} \setminus \mathfrak{U}_1, \end{cases}$$

where π_u is the ergodic distribution of the restriction of the chain X on E_u, $u \in \mathfrak{U}_1$.

1.3. Generalized potential operator

The property of the uniform ergodicity is closely related to the following notion.

DEFINITION 1.3. Let a kernel P have the stationary projector $\Pi \in \mathcal{B}$. A measure $\mu \in \mathfrak{M}$ is said to be the value $\mu = \nu R$ of a generalized potential R of the chain X if μ is a solution of the system

$$\mu(I - P) = \nu(I - \Pi), \qquad \mu\Pi = 0. \tag{1.8}$$

THEOREM 1.2. *A generalized potential R is well defined on a subspace*

$$\mathfrak{M}_0 = \mathfrak{M}(I - P + \Pi) \tag{1.9}$$

and it is a linear isomorphism between spaces $\mathfrak{M}_0(I - \Pi)$ and $\mathfrak{M}(I - \Pi)$ such that $(\mathfrak{M}_0\Pi)R = (\mathfrak{M}_0 R)\Pi = \{0\}$.

PROOF. To examine if Definition 1.3 is correct, it is sufficient to check that by Definition 1.2 from $\mu(I - P) = 0$, $\mu\Pi = 0$, follows that $\mu = \mu\Pi = 0$.

Let $\nu \in \mathfrak{M}_0$, that is, $\nu = \mu_1(I - P + \Pi)$, $\mu_1 \in \mathfrak{M}$. Then the measure $\mu = \mu_1(I - \Pi)$ meets (1.8), since $\mu\Pi = 0$ and $\mu_1(I - \Pi)(I - P) = \mu_1(I - P) = \mu_1(I - \Pi)(I - P + \Pi)$ according to (1.7) and $\mu_1(I - \Pi) \in \mathfrak{M}$ because of $\Pi \in \mathcal{B}$. So, for $\nu \in \mathfrak{M}_0$ the system (1.8) has a unique solution $\mu \in \mathfrak{M}$. The second assertion of the theorem follows from the equality $\mu = \mu_1(I - \Pi)$ obtained before. The equalities $\Pi R = R\Pi = 0$ are immediately derived from (1.8). If $\mu = \nu\Pi R$ then $\mu(I - P) = 0 = \mu\Pi$ and $\mu = 0$. Furthermore, if $\mu = \nu R\Pi$ then $\mu = \mu\Pi = 0$. The theorem is proved. □

THEOREM 1.3. *Assume that conditions* (M1)–(M3) *and* (P) *are satisfied. A chain X is uniformly ergodic with respect to a norm $\|\cdot\|$ if and only if it has a stationary projector $\Pi \in \mathcal{B}$ and one of the following equivalent (provided that $\Pi \in \mathcal{B}$) conditions holds true:*

(R) *the extension of the operator R on $\mathfrak{M}$ and $\|R\| \le r$ is bounded,*
(R1) *the operator $I - P + \Pi: \mathfrak{M} \to \mathfrak{M}$ has the bounded inverse operator;*
(R2) $\mathfrak{M}_0 = \mathfrak{M}$;
(R3) $\|\mu(I - \Pi)\| \le r\|\mu(I - P)\|$ *for all* $\mu \in \mathfrak{M}$.

PROOF. The equivalence of (R1) and (R2) directly follows from the definition of $\mathfrak{M}_0$: if the operator $S = (I - P + \Pi)^{-1}$ is bounded then $\mathfrak{M} = \mathfrak{M}S(I - P + \Pi) \subset \mathfrak{M}_0 \subset \mathfrak{M}$. On the other hand, if $\mathfrak{M}_0 = \mathfrak{M}$ then S is well defined on the whole $\mathfrak{M}$ and is closed. Indeed, $\mu(I - P + \Pi) = 0$ implies $\mu\Pi = 0$, $\mu = \mu P = \mu\Pi = 0$. Condition $\mathfrak{M}_0 = \mathfrak{M}$ yields that the equation $\mu(I - P + \Pi) = \nu$ has a solution for all $\nu \in \mathfrak{M}$ and, finally, from $\mu_n(I - P + \Pi) = \nu_n$, $\mu_n \to \mu$, $\nu_n \to \nu_0$, follows that $\mu_0(I - P + \Pi) = \nu_0$, so the operator S is closed. Therefore, under condition (R2) the operator S is bounded via the theorem on a closed graph, [see Kato (1966), Theorem 3.5.20].

The equivalence of conditions (R) and (R3) follows from Definition 1.3. Indeed, if $R \in \mathcal{B}$, $\|R\| \le r$, and $\mu = \nu R$, then $\mu\Pi = 0$ and

$$\|\mu(I - \Pi)\| = \|\mu\| = \|\nu R\| = \|\nu(I - \Pi)R\| \le r\|\nu(I - \Pi)\| = r\|\mu(I - P)\|.$$

Therefore, by Theorem 1.2, (R3) holds for $\mu \in \mathfrak{M}_0 R$ where $\mathfrak{M}_0 R = \mathfrak{M}(I - \Pi)$. It remains to observe that $\mu(I - P) = \mu(I - \Pi)(I - P)$ depends only on the measure $\mu(I - P)$. On the other hand, if condition (R3) is fulfilled, then the operator $(I - P)^{-1}$ is bounded on the Banach space $\mathfrak{M}^0 = \{\mu \in \mathfrak{M}: \mu\Pi = 0\}$ that is the closure of its domain, [see Kato (1966), Chapter 3.2 and Chapter 4.5]. Therefore, (1.8) immediately implies (R).

If condition (R1) holds true, then the system (1.8), that is equivalent to the following one $\mu(I - P + \Pi) = \nu(I - \Pi)$, $\mu\Pi = 0$, has a solution $\mu = \nu(I - \Pi)(I - P + \Pi)^{-1}$, $\|\mu\| \le c\|\nu\|$, that is, $R \in \mathcal{B}$. For $R \in \mathcal{B}$, we conclude from (1.8) that the measure $\mu = \nu R + \nu\Pi$ is a solution of the equation $\mu(I - P + \Pi) = \nu$ and $\|\mu\| \le c\|\nu\|$, that is, condition (R1) is fulfilled.

So, conditions (R)–(R3) are equivalent.

Assume that the chain X is uniformly ergodic. Then from the operator convergence $P^{(t)} \to \Pi$ we conclude that $\Pi \in \mathcal{B}$ and by Theorem 1.1 that Π is a stationary projector. For $n \ge 1$ consider the bounded operator $Q_n = I + n^{-1}\sum_{t=1}^{n}\left(P^{(t)} - \Pi\right)$. Taking into account (1.7), it is easy to check that

$$(I - P + \Pi)Q_n = Q_n(I - P + \Pi) = I - P^{(n)} + \Pi.$$

Choose n so that $\|P^{(n)} - \Pi\| < 1$. Then the operator $I - P^{(n)} + \Pi$ has the bounded inverse and commute with Q_n and $I - P + \Pi$. Dividing both sides of the equation obtained by its right-hand side, we arrive at the conclusion that the operator $(I - P + \Pi)^{-1} = Q_n\left(I - P^{(n)} + \Pi\right)$ is well defined and bounded. Therefore condition (R) is fulfilled as well as all other conditions of the theorem. □

We shall prove in Theorem 3.1 that a chain X is uniformly ergodic if $\Pi \in \mathcal{B}$ and $R \in \mathcal{B}$ and give the corresponding estimates for the convergence rate in Definition 1.1.

REMARK 1.4. The assumptions of Theorem 1.3 are necessary for the uniform ergodicity of a chain X even without conditions (M2) and (M3), since these are not used in the proof of the necessity. On the other hand, the sufficiency of these assumptions is essentially based on properties (M2) and (M3). This is illustrated in Example 1.1 below.

EXAMPLE 1.1. $E = [0,1)$, $\mathcal{E} = \mathcal{B}[0,1)$, and the space $\mathfrak{M}$ consists of finite measures such that the norm $\|\mu\| = \sum_{n\geq 0} |\int x^n \,\mu(dx)|$ is finite. Let the chain X be given by its transition kernel P defined as $Pf(x) = xf(0) + (1-x)f(x)$ for $f \in b\mathcal{B}$. For the chain X and the norm $\|\cdot\|$, conditions (M1) and (P) are satisfied and X has a unique invariant measure $\pi \in \mathfrak{M}$ that is concentrated at zero. According to Definition 1.3, one can easily derive from (1.8) that

$$\mu R(A) = \int_A x^{-1}\,\mu(dx) - \pi(A)\int_E x^{-1}\,\mu(dx)$$

for all measures μ with a support separated from zero. Note that this set is dense in $\mathfrak{M}$. By

$$\|\mu R\| = |\mu R(E)| + \sum_{n\geq 1}\left|\int x^n x^{-1}\,\mu(dx)\right| = \|\mu\|$$

we conclude that the generalized potential R is bounded, that is, condition (R) of Theorem 1.3 is satisfied.

However, the chain X is not uniformly ergodic with respect to the introduced norm $\|\cdot\|$. Indeed, the function $f(x) = x$ belongs to the dual space $\mathfrak{N}$ with a norm $\|f\| = \sup_{n,x} |f^{(n)}(x)|/n!$. The value of the corresponding Cesaro average is on f equal to $P^{(t)}f(x) = \left(1-(1-x)^t\right)/t$ and does not converge to $\Pi f(x) = 0$ in the norm $\mathfrak{N}$ which contradicts the uniform ergodicity.

It is clear that this contradiction appears because the norm is not monotone on $\mathfrak{M}$.

The following property of a potential shows that, for uniformly ergodic chains, it coincides analytically with the sum of an operator series.

THEOREM 1.4. *Let a chain X have the stationary projector $\Pi \in \mathcal{B}$, satisfies the condition* (P), *and $\|P^t\| = o(t)$ as $t \to \infty$. Then for all*

$$\mu \in \mathfrak{M}_{00} = \mathfrak{M}_0(I - P + \Pi) = \mathfrak{M}(I - P + \Pi)^2$$

the value μR coincides with the Cesaro sum of the series

$$\mu R = \sum_{t\geq 0}{}^{(c)} \mu\left(P^t - \Pi\right). \tag{1.10}$$

In particular, for uniformly ergodic chains, equation (1.10) *is valid for all $\mu \in \mathfrak{M}$.*

PROOF. It is sufficient to check that the series in (1.10) converges for $\mu \in \mathfrak{M}_{00}$ and its sum satisfies Definition 1.3.

Let $\mu = \nu(I - P + \Pi)^2$ for $\nu \in \mathfrak{M}$. Then, due to the conditions of the theorem, the Cesaro average of partial sums of the series (1.10) is equal to

$$\begin{aligned}\frac{1}{n}\sum_{t<n}\sum_{s<t}\nu(I-P+\Pi)^2(P^s-\Pi) &= \frac{1}{n}\sum_{t<n}\nu(I-P+\Pi)(I-P^t)\\ &= \nu(I-P+\Pi)+\frac{1}{n}(I-P^n+\Pi)\\ &= \nu(I-P)-\frac{1}{n}\nu(I-P^n)\to\nu(I-P)\end{aligned}$$

in the norm $\|\cdot\|$ as $n \to \infty$. This limit $\alpha = \nu(I - P)$ satisfies the relations $\alpha\Pi = \nu(I-P)\Pi = 0$, $\alpha(I-P) = \alpha(I-P+\Pi) = \nu(I-P)(I-P+\Pi) = \nu(I-P+\Pi)^2 - \nu\Pi = \mu - \nu\Pi = \mu(I - \Pi)$. Therefore $\alpha = \mu R$ for $\mu \in \mathfrak{M}_{00}$. Relation (1.10) is proved.

If the chain X is uniformly ergodic then $\|P^t\| = o(t)$ as $t \to \infty$, since $P^{(t)} \to \Pi$, and $\mathfrak{M}_{00} = \mathfrak{M}$ because the operator $(I - P + \Pi)$ has the bounded inverse. The theorem is completely proved. □

1.4. Resolvent of a transition kernel

Let us show that in Definition 1.1 the Cesaro averages may be replaced by the Abel averages. Simultaneously we prove some auxiliary results that will be used in further considerations.

The conditions (M1)–(M3) are supposed to be satisfied throughout below.

LEMMA 1.1. *Let μ, μ_1, μ_2 be some measures with $\mu_1 \ll \mu \ll \mu_2$, $\mu_i \in \mathfrak{M}$, $i = 1,2$, and let Q, Q_1, Q_2 be transition kernels with $Q_1 \leq Q \leq Q_2$, $Q_i \in \mathcal{B}$, $i = 1,2$. Then $\mu \in \mathfrak{M}$, $Q \in \mathcal{B}$, and*

$$\|\mu\| \leq \|\mu_1\| + \|\mu_2\|, \qquad \|Q\| \leq 2(\|Q_1\| + \|Q_2\|). \tag{1.11}$$

PROOF. Consider the Hahn decomposition $\mu = \mu^+ - \mu^-$, where $\mu^\pm \geq 0$, $\mu^+ \perp \mu^-$. From $\mu_1 \ll \mu \ll \mu_2$ follows that $\mu^+ \ll \mu_2^+$ and $\mu^- \ll \mu_1^-$, where $\mu_i^\pm$ are corresponding components of the Hahn decomposition for μ_i. Condition (M3) gives us $\mu_i^\pm \in \mathfrak{M}+$ and $\|\mu_i^\pm\| \leq \|\mu_i\|$. Consequently, condition (M2) implies $\mu_i^\pm \in \mathfrak{M}^+$, $\mu \in \mathfrak{M}$, and $\|\mu\| = \|\mu^+ - \mu^-\| \leq \|\mu^+\| + \|\mu^-\| \leq \|\mu_2^+\| + \|\mu_1^-\| \leq \|\mu_2\| + \|\mu_1\|$. So the first assertion of the lemma is valid.

Let $\mu \in \mathfrak{M}^+$. Then $\mu Q_1 \leq \mu q \leq \mu Q_2$ and $\mu Q \in \mathfrak{M}$, $\|\mu Q\| \leq \|\mu Q_1\| + \|\mu Q_2\|$. In the general case, due to condition (M3), we expand $\mu \in \mathfrak{M}$ as $\mu = \mu^+ - \mu^-$, $\mu^+ \perp \mu^-$, where $\mu^\pm \in \mathfrak{M}$ and note that $\|\mu^\pm\| \leq \|\mu\|$, whence $\|\mu^+\| + \|\mu^-\| \leq 2\|\mu\|$. Hence

$$\begin{aligned}\|\mu Q\| &\leq \|\mu^+ Q\| + \|\mu^- Q\| \leq \|\mu^+ Q_1\| + \|\mu^+ Q_2\| + \|\mu^- Q_1\| + \|\mu^- Q_2\|\\ &\leq (\|\mu^+\| + \|\mu^-\|) \leq (\|Q_1\| + \|Q_2\|) \leq 2\|\mu\|\,(\|Q_1\| + \|Q_2\|)\,.\end{aligned}$$

Since μ is arbitrary, the last inequality implies (1.11). □

In what follows, the analyticity of operator-valued functions is understood in the strong sense.

LEMMA 1.2. *Let the kernel Q be nonnegative and $Q \leq P$, where the stochastic kernel P is such that $\|P^t\| = o(t)$ as $t \to \infty$ (for example, a chain with transition probabilities P may be uniformly ergodic). Then the resolvent $Q_z = (z - Q)^{-1} \in \mathcal{B}$ is analytical in the domain $D = \{z: |z| > 1\}$.*

Note that a chain with transition probabilities P that satisfy conditions of Lemma 1.2 is uniformly ergodic.

PROOF. By Theorem 10.19 [Rudin (1973)] it is sufficient to show that the spectral radius $\rho(Q) = \lim\left(\|Q^n\|\right)^{1/n}$ of the operator Q does not exceed 1, since the inclusion $Q \in \mathcal{B}$ follows from Lemma 1.1. To do this note that, by Lemma 1.1, $\rho(Q) \leq \lim(2\|P^n\|)^{1/n} = \rho(P)$ and $\rho(P) \leq \lim(Cn)^{1/n} = 1$ by the assumptions of the lemma. This is precisely what had to be proved. □

Assume that the kernel P of a chain X satisfies the condition (P). Consider an operator-valued function

$$R_z = (1 - z)(I - zP)^{-1} \tag{1.12}$$

defined for z^{-1} from the resolvent domain of the operator P.

THEOREM 1.5. *A chain X is uniformly ergodic in the norm $\|\cdot\|$ if and only if there exists a stochastic kernel $\Pi \in \mathcal{B}$ such that $\|R_z - \Pi\| \to 0$ as $z \to 1$ along some path within the unit circle $\{z: |z| < 1\}$. Moreover, in this case Π coincides with the stationary projector of the kernel P.*

PROOF. Let $R_z \to \Pi$ in the operator topology. From Definition (1.12) it is easily seen that

$$PR_z = R_zP = z^{-1}(R_z + (z - 1)I). \tag{1.13}$$

Approaching the limit in (1.13) as $z \to 1$ we obtain $\Pi P = \Pi = P\Pi = \Pi R_z = \lim \Pi R_z = \Pi^2$ and $\mu = \mu R_z = \lim \mu R_z = \mu\Pi$ with $\mu = \mu P$, that is, Π is the stationary projector for P. Therefore $R_z\Pi = \Pi R_z = \Pi$.

It is easy to check by using (1.13) and (1.7) that

$$z(R_z - z\Pi)(I - P + \Pi) = (1 - z)(I - R_z + z\Pi), \tag{1.14}$$

where all factors are commuting. Since $R_z \to \Pi \in \mathcal{B}$, there exists $z \neq 1$ such that $\|R_z - z\Pi\| < 1$. Then $(I - R_z + z\Pi)^{-1} \in \mathcal{B}$ and from (1.14) we derive that the bounded operator $z(1-z)^{-1}(R_z - z\Pi)(I - R_z + z\Pi)^{-1}$ coincides with $(I - P + \Pi)^{-1}$. Therefore, by Theorem 1.3 the chain X is uniformly ergodic.

To prove the necessity, we use again Theorem 1.3. The operator $(I - P + \Pi)$, for $\Pi \in \mathcal{B}$ to be the stationary projector of P, has the bounded inverse by (R1). Since $(I - zP + z\Pi) - (I - P + \Pi) = (1 - z)(P - \Pi) \to 0$ as $z \to 1$, $(I - zP + z\Pi)^{-1} \in \mathcal{B}$ for $z - 1$ small enough because it has the bounded inverse. Besides, the norm of this operator is uniformly bounded in some neighborhood of the unity. It remains to note that (1.14) yields $(R_z - \Pi)(I - zP + z\Pi) = (1 - z)(I - \Pi)$. Multiplying this identity by $(I - zP + z\Pi)^{-1}$ we make sure that $\|R_z - \Pi\| = O(|z - 1|)$ as $z \to 1$. The theorem is proved. □

1.5. Strongly stable chains

It is assumed in this section that a chain X has a unique invariant probability measure $\pi \in \mathfrak{M}$ and conditions (M1)–(M3) and (P) are satisfied.

DEFINITION 1.4.. A Markov chain X is said to be strongly stable in the norm $\|\cdot\|$ if any stochastic kernel $Q \in \mathcal{B}$ has in some neighborhood $\{Q \in \mathcal{B}: \|Q-P\| < \varepsilon\}$ a unique invariant probability measure $\nu \in \mathfrak{M}$ with $\|\nu - \pi\| \to 0$ as $\|Q - P\| \to 0$ uniformly in this neighborhood.

THEOREM 1.6. *A chain X is strongly stable in the norm $\|\cdot\|$ if and only if it is uniformly ergodic in the same norm.*

PROOF. Let the chain X be strongly stable. To prove the uniform ergodicity of X we use Theorem 1.5. Choose $\alpha \in \mathfrak{M}^+$ such that $\|\alpha\| \le k^{-1}$ for the constant k from condition (M1). Thus $\alpha\mathbb{I} \le \|\alpha\| \cdot \|\mathbb{I}\| \le k^{-1}k = 1$. For $u \in (0,1)$, define a stochastic kernel

$$Q = uP + (1-u)\mathbb{I} \circ (\alpha + (1 - \alpha\mathbb{I})\pi).$$

Condition (P) and inclusion $\pi \in \mathfrak{M}$ yield $Q \in \mathcal{B}$ because of $\alpha \in \mathfrak{M}$. We write down

$$\begin{aligned}\|Q-P\| &= (1-u)\|P - \mathbb{I} \circ (\alpha + (1-\alpha 1)\pi\| \\ &\le (1-u)(\|P\| + k\|\alpha\| + k(1-\alpha 1)\|\pi\|) \le (1-u)(\|P\| + \mathbb{I} + k\|\pi\|).\end{aligned} \tag{1.15}$$

Therefore, $\|Q - P\| \to 0$ as $u \to 1$ uniformly on $\alpha \in \{\alpha: \|\alpha\| \le k^{-1}\}$. Thus, for sufficiently small $1-u$, for the kernel Q, there exists a unique invariant probability measure ν with $\|\nu - \pi\| \to 0$ as $u \to 1$.

The measure ν meets the equation $\nu = \nu Q$, whence

$$\nu(I - uP) = (1-u)(\alpha + (1-\alpha 1)\pi),$$

since $\nu 1 = 1$. Therefore $\nu = \alpha R_u + (1-u)\pi R_u = \alpha(R_u - \Pi) + \pi$ with the kernel $R_u \in \mathcal{B}$ defined by (1.12) and $\Pi = \mathbb{I} \circ \pi$ to be the stationary projector of P.

By the strong stability condition, for any $\varepsilon > 0$, there exists $\delta > 0$ such that $\|\nu - \pi\| = \|\alpha(R_u - \Pi)\| < \varepsilon$ for all $\alpha \in \mathfrak{M}^+$ with $\|\alpha\| \le k^{-1}$ provided that $\|Q - P\| < \delta$. Using (1.15) and the linearity of $R_u - \Pi$ we conclude from the last assertion that

$$\|\alpha(R_u - \Pi)\| \le \varepsilon k \|\alpha\| \tag{1.16}$$

for all $\alpha \in \mathfrak{M}^+$ and all $0 < 1-u < \delta/c$, where C does not depend on α.

Let $\mu \in \mathfrak{M}$. Expand μ as $\mu = \mu_1 - \mu_2$ with $\mu_i \in \mathfrak{M}^+$ and $\mu_1 \perp \mu_2$. Condition (M3) and (1.16) yield

$$\|\mu(R_u - \Pi)\| \le \|\mu_1(R_u - \Pi)\| + \|\mu_2(R_u - \Pi)\| \le \varepsilon k(\|\mu_1\| + \|\mu_2\|) \le 2\varepsilon k\|\mu\| \tag{1.17}$$

for all $0 < 1-u < \delta/c$. Inequality (1.17) is valid uniformly by $\mu \in \mathfrak{M}$, and thus it implies that the operator convergence $R_u \to \Pi$ holds true as $u \to 1$. By Theorem 1.5 the latter implies the uniform ergodicity of X. □

The strong stability of a uniformly ergodic chain X will be established in Theorem 3.3 below, along with estimates of stability.

THEOREM 1.7. *Let a chain X have a unique invariant probability measure. Then the property of the uniform ergodicity (and of the strong stability) of the chain X with respect to the norm $\|\cdot\|$ is preserved under small (in a corresponding operator norm) perturbations of its transition kernel P, that is, the uniform ergodicity of the chain X in the norm $\|\cdot\|$ implies that any stochastic kernel Q from an operator neighborhood $\{Q : \|Q - P\| < \varepsilon\}$ corresponds to a uniformly ergodic and strong stable in this norm Markov chain.*

PROOF. If the chain X is uniformly ergodic, then by Theorem 1.3 the operator $I - P + \Pi$ has the bounded inverse and $\Pi \in \mathcal{B}$, and by Theorem 1.6 for any $Q \in \mathcal{B}$ in a neighborhood $\{Q : \|Q - P\| < \varepsilon\}$ there exists a unique invariant measure $\nu \in \mathfrak{M}$ with $\|\nu - \pi\| \to 0$. Recall that (see Remark 1.2) the stationary projector N of a chain with the kernel Q is of the form $N = \mathbb{1} \circ \nu$. It remains to note that $(I - Q + N) - (I - P + \Pi) = P - Q + \mathbb{1} \circ (\nu - \pi) \to 0$ as $\|Q - P\| \to 0$ in the operator topology, since $\mathbb{1} \in \mathfrak{N}$ and $\|\nu - \pi\| \to 0$. Therefore, for a sufficiently small $\|Q - P\|$, the operator $I - Q + N$ also has the bounded inverse and the assertion of the theorem follows from Theorem 1.3. □

CHAPTER II

CRITERIA FOR THE UNIFORM ERGODICITY AND ITS PROPERTIES

The notion of uniform ergodicity introduced above is connected with the asymptotics of Cesaro's (or Abel's) averages of the transition probabilities. In this Chapter, we investigate the asymptotics of powers of the transition kernel under an additional mixing condition, since it describes the whole spectrum of the transition operator. We define the notion of a period of a chain and corresponding cyclic subclasses. Spectral properties of the operator P are studied, too. We also obtain a criterion for the uniform ergodicity generalizing the well known Doeblin condition and the uniform mixing condition.

2.1. Mixing property

It what follows we assume that conditions (M1)–(M3) are satisfied for the space $\mathfrak{M}$ and that a Markov chain $X = (X_t, t \geq 0)$ on $(E, \mathcal{E})$ with the kernel $P(x, A)$ satisfies the condition (P) and has a unique invariant probability measure $\pi \in \mathfrak{M}$. Then the corresponding stationary projector is equal to $\Pi = \mathbb{I} \circ \pi$ (see Remark 1.2) and, moreover, $\Pi \in \mathcal{B}$.

Besides, the following mixing condition is used in the sequel

(C) there exist a natural n, a measure $\alpha \in m\mathcal{E}^+$, and a function $h \in b\mathcal{E}^+$ such that $\pi h > 0$, $\alpha(E) = 1$, $\alpha h > 0$, and the kernel

$$T(x, A) = P^n(x, A) - h(x)\alpha(A)$$

is nonnegative.

REMARK 2.1. Condition (C) is fulfilled for all Harris chains X where $\alpha(A) = \pi(A \cap C)/\pi(C)$, $h(x) = \varepsilon\pi(C)\mathbb{I}_{\{x \in C\}}$, $\varepsilon > 0$, and the set C are such that the density $p_n(x, y)$ of an absolutely continuous component of the measure $P^n(x, \cdot)$ with respect to $\pi(\cdot)$ is greater than ε on the set $(x, y) \in C \times C$. Such ε and C exist as stated in Lemma 6.1.1 in Revuz (1975).

2.2. A period of a chain. Cyclic classes

To investigate the asymptotic behaviour of powers P^t we introduce the following definition.

DEFINITION 2.1. The number

$$d(X) = \sup_{n \geq 1} \dim\{\, \mu \in \mathfrak{M}: \mu = \mu P^n \,\} \tag{2.1}$$

is said to be a period of a chain X.

If $d(X) = 1$ then the chain X is said to be aperiodic.

THEOREM 2.1. *Let a chain X be uniformly ergodic with respect to the norm $\|\cdot\|$. Then it has a finite period $d(X) = d$ and there exists a partition $E = \bigcup_{i=0}^{d} E_i$, $E_i \in \mathcal{E}$, $E_i \cap E_j = \varnothing$, $i \neq j$, such that*

(a) *for any $1 \le i \le d$ the restriction P_i of the kernel P^d on the subspace $(E_i, E_i \cap \mathcal{E})$ is a stochastic transition kernel and corresponds to an aperiodic uniformly ergodic Markov chain with respect to the induced norm*

$$\|\mu(\cdot)\|_i = \|\mu(E_i \cap (\cdot))\|;$$

(b) *the set E_0 is transient, that is, $\pi(E_0) = 0$ and $\left\|\mu P^{(t)}\right\|_0 \to 0$ as $t \to \infty$ uniformly on $\mu \in \{\|\mu\| \le 1\}$.*

The proof of the theorem is based on the following auxiliary results.

LEMMA 2.1. *Let conditions* (M1)–(M3) *be satisfied and let a substochastic kernel Q belong to $\mathcal{B}$. Then the space $M = \{\mu \in \mathfrak{M}: \mu = \mu Q\}$ is a substructure in $\mathfrak{M}$ generated by the cone $M^+ = M \cap \mathfrak{M}^+$, $M = M^+ - M^+$.*

PROOF. Let $\mu \in \mathfrak{M}$. We represent μ as $\mu = \mu_1 - \mu_2$ with $\mu_i \in \mathfrak{M}^+$, $\mu_1 \perp \mu_2$. Then, for some $E' \in \mathcal{E}$, $\mu_1(E \setminus E') = \mu_2(E') = 0$. Therefore, for all $A \in \mathcal{E}$,

$$\mu_1(A) = \mu_1(A \cap E') = (\mu_1 - \mu_2)(A \cap E') = \mu(Q \cap E'),$$
$$\mu Q(A \cap E') \le \mu_1 Q(A \cap E') \le \mu_1 Q(A).$$

Since $\mu_1(E) \ge \mu_1 Q(E)$, the inequality $\mu_1 \le \mu_1 Q$ implies $\mu_1 = \mu_1 Q$, $\mu_1 \in M^+$. Then $\mu_2 = \mu + \mu_1 \in M^+$ and consequently $M = M^+ - M^+$. The lemma is proved. □

LEMMA 2.2. *Let the stochastic kernel $P \in \mathcal{B}$ have a unique invariant probability measure π, $\pi \in \mathfrak{M}$, and $n \ge 1$ be positive integer. Put $M_n = \{\mu \in \mathfrak{M}: \mu = \mu P^n\}$, $d_n = \dim M_n$. Then $d_n \le n$.*

PROOF. By Lemma 2.1 we have $d_n = \dim M_n^+$ where $M_n^+ = M_n \cap \mathfrak{M}^+$. Let $\alpha \in M_n^+$. Then $\alpha(I - P^n) = 0 = \alpha\left(I + \cdots + P^{n-1}\right)(I - P)$, that is, the measure $\alpha\left(I + \cdots + P^{n-1}\right)$ is invariant for P. By assumptions, this measure is proportional to π, whence

$$\alpha\left(I + \cdots + P^{n-1}\right) = n\pi \tag{2.2}$$

for all $\alpha \in M_n^+$ with $\alpha \mathbb{I} = \mathbb{I}$. It follows from (2.2), in particular, that $\alpha \le n\pi$ for all such α.

Using in M_n^+ the orthogonalization procedure, [see Birkhoff (1967)], we construct a sequence of mutually orthogonal probability measures $\{\alpha_i, i \le d_n\}$, $d_n \le \infty$, $\alpha_i \in M_n^+$, $\alpha_i \mathbb{I} = 1$. Let E_i be the support of α_i, that is, $\alpha_i(E_i) = 1$, and $E_i \cap E_j = \varnothing$ for $i \neq j$. Then

$$d_n = \sum \alpha_i(E_i) \le \sum n\pi(E_i) = n\pi\left(\bigcup E_i\right) \le n,$$

and the proof is completed. □

LEMMA 2.3. *Let the stochastic kernel $P \in \mathcal{B}$ have the unique invariant probability measure π, $\pi \in \mathfrak{M}$, and let $n \geq 1$ be an integer and $d = d_n = \dim\{\mu \in \mathfrak{M}: \mu = \mu P^n\}$. Then there exists a partition $E = \bigcup_{i=0}^d E_i$, $E_i \in \mathcal{E}$, $E_i \cap E_j = \varnothing$, $i \neq j$, such that*

(1) $P(x, E_{i+1}) = 1$ *for* $x \in E_i$, $1 \leq i < d$;

(2) $P(x, E_1) = 1$ *for* $x \in E_d$;

(3) *there exists a collection* $\{\pi_1, \dots, \pi_d\}$ *of probability measures with* $\pi(E_i) = 1$ *such that*

$$\frac{1}{d}(\pi_1 + \cdots + \pi_d) = \pi \tag{2.3}$$

and $\pi_i P = \pi_{i+1}$ *for* $1 \leq i < d$, $\pi_d P = \pi_1$;

(4) $\pi(E_0) = 0$.

PROOF. By Lemma 2.2 we have $d \leq n$. Thus the space $M_n = \{\mu \in \mathfrak{M}: \mu = \mu P^n\}$ is finite dimensional. Since $M_n = M_n^+ - M_n^+$ by Lemma 2.1, there exists in M_n a finite basis $\{\pi_1, \dots, \pi_d\}$ of mutually orthogonal measures $\pi_i \in M_n^+$, $\pi_i(E) = 1$.

Let us decompose the measure $\pi_i P \in M_n^+$ with respect to this basis: $\pi_i P = \sum \lambda_{ij}\pi_i$. Then $\pi_i P^t = \sum \lambda_{ij}\pi_i P^{t-1} = \sum \lambda_{ij} \sum \lambda_{jk}\pi_k P^{t-2} = \cdots = \sum \lambda_{ij}^{(t)}\pi_j$ for any $t > 1$, where the matrix $\Lambda^{(t)} = \left(\lambda_{ij}^{(t)}, 1 \leq i, j \leq d\right)$ coincides with the t-th power Λ^t of the matrix $\Lambda = (\lambda_{ij}, 1 \leq i, j \leq d)$. Since $\{\pi_1, \dots, \pi_d\}$ is a basis in M_n^+ and $\pi_i P^n = \pi_i$, $\Lambda^n = I_d$ where I_d is the unit $d \times d$ matrix. It is easily seen that the matrix Λ is stochastic because of $1 = \pi_i P \mathbb{I} = \sum \lambda_{ij}\pi_j \mathbb{I} = \sum \lambda_{ij}$ and $\lambda_{ij} \geq 0$, since $\pi_i P \geq 0$ and the measures π_i are orthogonal. Therefore, we get from $\Lambda^n = I_d$ that $\lambda_{ij} \in \{0, 1\}$, Λ is a permutator, and n is divisible by d. We arrange in an appropriate order the elements of the basis $\{\pi_1, \dots, \pi_d\}$ and obtain from the definition of Λ that $\pi_i P = \pi_{i+1}$ for $1 \leq i \leq d$, $\pi_d P = \pi_1$. Moreover, (2.2) yields $n\pi = \pi_1(I + \cdots + P^{n-1}) = (\pi_1 + \cdots + \pi_d) + \cdots + (\pi_1 + \cdots + \pi_d) = (\pi_1 + \cdots + \pi_d)n/d$, so equation (2.3) holds true.

Let E_i' be the support of the measure π_i, that is, $E_i' \in \mathcal{E}$, $\pi_i(E \setminus E_i') = 0$, $E_i' \cap E_j' = \varnothing$, $i \neq j$. Define successively $E_0 = E_d'$ and $E_{t+1} = \{x \in E_{t+1-d}: P(x, E_t) = 1\}$, $t \geq 0$, where by the definition we put $E_l = e'_{-l}$ for $-d < l \leq -1$. Since the kernel P is regular, $E_t \in \mathcal{E}$. Furthermore, $E_t \in \mathcal{E}$, $t \geq 1$, and $E_t \cap E_l = \varnothing$ for $t \neq l \mod d$, since the sets E_l' are pairwise disjoint.

Besides, $\pi_l(E_t) = 1$ for $t + l = 0 \mod d$. Here and in the sequel we set by the definition $\pi_{t+d} = \pi_t$ for all t. Indeed, for $t = -d$ the equation above is valid. Let it be valid for some t and $l = -t \mod d$. We derive from the inequality

$$1 = \pi_l(E_t) = \int \pi_{l-1}(dx)P(x, E_t) \leq \pi_{l-1}(E_{t+1}) + \int_{E \setminus E_{t+1}} \pi_{l-1}(dx)P(x, E_t)$$

that

$$\int_{E \setminus E_{t+1}} (1 - P(x, E_t))\pi_{l-1}(dx) = 0. \tag{2.4}$$

Since $P(x, E_t) < 1$ on $E \setminus E_{t+1}$, we conclude from (2.4) that $\pi_{l-1}(E \setminus E_{t+1}) = 0$. Thus $\pi_{l-1}(E_{t+1}) = 1$ and the proof of the equality desired is completed by induction.

Let $\overline{E}_i = \bigcap_{t\geq 1} E_{td-i}$, $1 \leq i \leq d$. Then $\overline{E}_i \in \mathcal{E}$, since $E_t \in \mathcal{E}$. Moreover, $\pi_i\left(\overline{E}_i\right) = \lim_{t\to\infty} \pi_i(E_{td-i}) = 1$, since the measures π_i are countable additive. By the definition of E_t we have $P\left(x, \overline{E}_{t+1}\right) = 1$ for all $x \in \overline{E}_t$. Moreover, the sets $\overline{E}_i \subset E_i'$ are disjoint. It remains to note that

$$\pi\left(E \setminus \bigcup_{i=1}^{d} E_i\right) = 1 - \sum_{i=1}^{d} \pi\left(\overline{E}_i\right) = 1 - \frac{1}{d}\sum_i\sum_j \pi_j\left(\overline{E}_i\right)$$
$$= \frac{1}{d}\sum_i \pi_i\left(\overline{E}_i\right) = 1 - 1 = 0,$$

so that the sets $\overline{E}_i$, $1 \leq i \leq d$ and $\overline{E}_0 = E \setminus \bigcup \overline{E}_i$ as desired. The proof is completed. □

LEMMA 2.4. *Let a chain X with the kernel P be uniformly ergodic with respect to the norm $\|\cdot\|$. Then, for any $n \geq 1$, the unity is isolated in the spectrum of the operator $P^n: \mathfrak{M} \to \mathfrak{M}$ and the operator $I - P^n$ is of the Fredholm type, [see Kato* (1966), *Chapter* 4.5].

PROOF. The stationary projector Π of the kernel P, by Theorem 1.3, is of the form $\Pi = \mathbb{I} \circ \pi$, belongs to $\mathcal{B}$, and the operator $S = (I - P + \Pi)^{-1}$ is bounded on $\mathfrak{M}$. Therefore a continuation of S on the second dual space $\mathfrak{M}^{**}$ is bounded, too, [see Rudin (1973), Theorem 4.10]. The space $\mathfrak{M}_a = \mathfrak{M}^{**}$ that is dual to $\mathfrak{N} \subset f\mathcal{E}$ consists of finite-additive measures μ on $\mathcal{E}$ such that $\sup(|\mu f|, \|f\| \leq 1) = \|\mu\| < \infty$, [see Rudin (1973), Chapter 4.5]. Consider the equation $\mu = \mu P$ for $\mu \in \mathfrak{M}_a$, $\mu\mathbb{I} = 1$. It is equivalent to the equation $\mu(I - P + \Pi) = \mu\Pi = \pi$, whence $\mu = \pi S = \pi$ in view of the boundedness of S on $\mathfrak{M}_a$. Therefore, the kernel P has no finite-additive invariant measures on $\mathfrak{M}_a \setminus \mathfrak{M}$. The kernel P^n possesses the same property for all $n \geq 1$. Indeed, if $\mu = \mu P^n$ for some $n \geq 1$ and $\mu \in \mathfrak{M}_a \setminus \mathfrak{M}$ then the measure $\mu\left(I + \cdots + P^{n-1}\right)$ is invariant for P and has a component μ, $\mu \in \mathfrak{M}_a \setminus \mathfrak{M}$. Since each purely finite-additive measure is orthogonal to any countable-additive measure and, according to Lemma 2.1, the space $\{\mu \in \mathfrak{M}: \mu = \mu P\}$ is a substructure in $\mathfrak{M}$, the purely finite-additive component of the measure $\mu \in \mathfrak{M}_a \setminus \mathfrak{M}$ is invariant for P and not equal to zero. This contradicts the part of the lemma proved above. Therefore, the operator $I - P^n$ has no annihilator in $\mathfrak{M}_a \setminus \mathfrak{M}$. Since the space of annihilators of the operator $I - P^n$ in $\mathfrak{M}$ is of finite dimension (by Lemma 2.2 its dimension does not exceed n), the corresponding subspace of $\mathfrak{M}_a$ also is of a finite dimension.

Hence, the space $(I - P^n)\mathfrak{N}$ is closed in $\mathfrak{N}$ because, by Theorem 4.5.31 from Kato (1966), its defect is equal to the dimension of the space $\{\mu \in \mathfrak{N}^*: \mu = \mu P^n\}$. The latter is finite because, by the definition, $\mathfrak{N}^* = \mathfrak{M}^{**} = \mathfrak{M}_a$. Thus the operator $I - P^n: \mathfrak{N} \to \mathfrak{N}$ is of the semi-Fredholm type, [see Kato (1966), Chapter 4.5]. By Theorem 4.5.31 in Kato (1966) the operator $zI - P^n$ is also of the semi-Fredholm type for $|z - 1| > 0$ small enough, and its index and defect does not depend on z. However the operator $zI - P^n$ is invertible for $|z| > 1$ on $\mathfrak{M}$ by Lemma 1.2 (and thus on $\mathfrak{N}$ by Theorem 3.5.30 in Kato (1966)). Therefore these index and defect are equal to zero and $(zI - P^n)^{-1} \in \mathcal{B}$ for all sufficiently small $|z - 1| > 0$. Thus the unity is isolated in the spectrum of the operator P^n. Now from the stability of the index of an operator for bounded small perturbations, [see Theorem 4.5.17 in Kato (1966)], we obtain that the operator $I - P^n$ is of the Fredholm type because so is $zI - P^n$. Hence the lemma is proved. □

PROOF OF THEOREM 2.1. We show first that the period $d(X)$ of the chain X defined by (2.1) is finite. According to Lemma 2.4, for all n, the unity is isolated in the spectrum $\sigma(P^n)$ of the operator $P^n: \mathfrak{M} \to \mathfrak{M}$. By the theorem on a spectrum transformation [see Theorem 7.3.19 in Dunford and Schvartz (1953)] $\sigma(P^n) = \{z^n, z \in \sigma(P)\}$. Therefore all the roots of the unity are isolated in the spectrum $\sigma(P)$ of the operator P. Let $d = \dim\{\mu \in \mathfrak{M}: \mu = \mu P^n\}$ and ω be the simple d-fold root of the unity. Then $\omega^k \in \sigma(P)$ for all $1 \le k \le d$. Indeed, by the definition of the measures π_i given in Lemma 2.3 it follows that $\mu_k \omega^k = \mu_k P$ where $\mu_k = \sum_{j=1}^{d} \omega^{jk} \pi_j$.

Since $\sigma(P) \cap \{z: 0 < |z-1| < \varepsilon\} = \varnothing$ for some $\varepsilon > 0$, the spectrum $\sigma(P)$ does not contain a simple root ω_k of a degree $k > l = [\pi/\arcsin(\varepsilon/2)] + 1$ because otherwise $|\omega_k - 1| < \varepsilon$. Therefore, $\sigma(P)$ does not contain roots of the unity with degrees greater than l.

The dimension of the space $\{\mu \in \mathfrak{M}: \mu = \mu P^l\}$, by Lemma 2.2, is finite (even does not exceed l) and by the spectrum transformation theorem this dimension coincides with the dimension of a spectral subspace of the operator P which is associate with a spectral set $\{\omega: \omega^l = 1\}$. Thus the dimension of the space $\{\mu \in \mathfrak{M}: \mu = \mu P^n\}$ does not exceed l for each $n \ge 1$, since this space corresponds to the spectral subspace of the spectral set $\{\omega: \omega^n = 1\} \cap \sigma(P) \subset \{\omega: \omega^l = 1\}$. Therefore, $d(X) \le l$ by the definition (2.1) and the chain X has a finite period. Consider the sets E_i and measures π_i defined in Lemma 2.3 for $d = d(X)$ and $n = d$. Then $P^d(x, E_i) = 1$ for $x \in E_i$, $1 \le i \le d$, and the kernel $P_i(x, A) = P^d(x, A \cap E_i)$ corresponds to a Markov chain on $(E_i, E_i \cap \mathcal{E})$. This chain is aperiodic, otherwise the period of X (that coincides with the maximal dimension allowed in (2.1)) would be greater than $d(X) = d$.

Note that the equation $\alpha = \alpha P_i$, $\alpha \in \mathfrak{M}$, $\alpha 1 = 1$, has the unique solution $\alpha = \pi_i$. Otherwise the dimension of the invariant subspace for P is greater than d. Moreover, the estimate $\pi_i \le d\pi$, that follows from (2.2), and the inclusion $\pi \in \mathfrak{M}$ give us $\pi_i \in \mathfrak{M}$ in view of condition (M2). Hence, as in Theorem 1.3, the operator $I - P_i + \Pi_i$ has no annihilators in the space $\mathfrak{M}_i = \{\mu \in \mathfrak{M}: |\mu|(E \setminus E_i) = 0\}$, where $\Pi_i = \mathbb{I} \circ \pi_i$ is the stationary projector of P_i. Besides $\Pi_i \in \mathcal{B}$ because $\pi_i \in \mathfrak{M}$.

The restriction of the operator $I - P^d$ on the space $\mathfrak{M}_i$ coincides with $I - P_i$ by the definition of P_i. Thus by Lemma 2.4 the operator $I - P_i: \mathfrak{M}_i \to \mathfrak{M}_i$ is of the Fredholm type as well as $I - P^d$. Because of the finite dimension of Π_i the operator $I - P_i + \Pi_i$ is also of the Fredholm type by Theorem 4.5.6 in Kato (1966). It was already proved that this operator has an empty space of annihilators. Therefore by Theorem 1.3 the inverse operator $(I - P_i + \Pi_i)^{-1}$ is bounded on $\mathfrak{M}_i$ and the chain with the kernel P_i is uniformly ergodic in the norm that is nothing other than $\|\cdot\|$ on the space $\mathfrak{M}_i$.

It remains to note that $\mu\Pi(A) = (\mu\mathbb{I})\pi(A) = 0$ for all $\mu \in \mathfrak{M}$ and $A \in E_0 \cap \mathcal{E}$. Thus $\|\nu - \mu\Pi\|_0 = \|(\nu - \mu\Pi)(E_0 \cap (\cdot))\| = \|\nu\|_0$ for all measures ν. Therefore, by Definition 1.1, $\|\mu P^{(t)}\|_0 = \|\mu\left(P^{(t)} - \Pi\right)_0 \le \|\mu\left(P^{(t)} - \Pi\right) \to 0$ uniformly on

$\{\|\mu\| \leq 1\}$ as $t \to \infty$. The theorem is proved. □

2.3. The asymptotics of powers of a transition kernel

Define the following ergodicity index which depends only on a kernel P

$$\Lambda_n^d(P) = \sup\left(\|\mu P^n\|, \|\mu\| \leq 1, \mu(E) = 0, \mu \in \mathfrak{M}P^{(d)}\right). \tag{2.5}$$

THEOREM 2.2. *Let condition* (C) *hold true. A chain X with the kernel P is uniformly ergodic with respect to a norm $\|\cdot\|$ if and only if for some $d \geq 1$ one of the following equivalent conditions is valid*:

(A1) $\Lambda_n^d(P) < 1$ *for some* $n \geq 1$;

(A2) $\Lambda_n^d(P) \to 0$ *as* $t \to \infty$;

(A3) *there exists a compact operator $K \in \mathcal{B}$ such that*

$$\|P^n - K\| < 1$$

for some $n \geq 1$;

(A4) *for some $0 < \rho < 1$, the spectrum of the operator $P\colon \mathfrak{M} \to \mathfrak{M}$ contains in the domain $\{z\colon |z| > \rho\}$ only those roots of the unity that are simple poles of the resolvent and are of a degree that does not exceed d*;

(A5) $\|P^{(d)}(P^t - \Pi)\| = O(\rho^t)$ *as* $t \to \infty$ *for some* $0 < \rho < 1$;

(A6) *there exists an orthogonal projector $\Pi_0 \in \mathcal{B}$ of a dimension that does not exceed d and such that $\|P^{td} - \Pi_0\| = O(\rho^t)$ as $t \to \infty$ for some $0 < \rho < 1$.*

Moreover, the minimal d satisfying the conditions above coincides with the period $d(X)$ in (2.1) *and the dimension of Π_0 is equal to $d(X)$.*

REMARK 2.2. The conditions (A1)–(A6) are equivalent even in the case where assumption (C) is not valid. Each of them implies the uniform ergodicity of the chain X. Note that the definition of the uniform ergodicity is connected only with a part of the spectrum of the operator P in a small neighborhood of the unity, whereas (A1)–(A6) describe this spectrum in the whole unit circle. The equivalence stated in the theorem above is essentially connected with the mixing condition (C) and with properties (M1)–(M3) of the space $\mathfrak{M}$.

The last sentence is clarified in the following example.

EXAMPLE 2.1. Let $(\xi_t, t \geq 1)$ be a sequence of independent random variables with values in $E = [0,1)$ and with a common distribution $F(B) = \mathsf{P}(\xi_t \in B)$, $B \in \mathcal{B}[0,1)$. Define the chain $X_t = X_0 + \xi_1 + \cdots + \xi_t \mod 1$ on E with the transition kernel $P(x, B) = F(B - x)$ and with the stationary projector $\Pi(x, B) = L(B)$, where L is the Lebesque measure. The projector is defined uniquely, provided that the distribution F is non-lattice, that is, $|\omega_n| < 1$ for $n \neq 0$ where $\omega_n = \mathsf{E}\exp\{2\pi i n \xi_1\}$ are Fourier coefficients.

Denote by $\mathfrak{M}$ the space of all signed measures on E with square integrable densities and by $\|\mu\|$ the $L_2(E)$-norm of the density of the measure μ. Let also $\varphi_n(x) = \exp\{2\pi i n x\}$ and $\mu\varphi_n$ be the integral of the function φ_n over the measure μ.

Assume that the set of limit points of the sequence ω_n is dense on the unit circle (for example, this is the case if F has two irrational linearly independent points of growth). Then the chain X is not uniformly ergodic with respect to the norm $\|\cdot\|$ and therefore does not satisfy the condition (C). At the same time we have $\|\mu(P^t - \Pi)\| \to 0$ as $t \to \infty$ for all $\mu \in \mathfrak{M}$.

For $\varepsilon > 0$, we introduce the set $\mathcal{N}_\varepsilon = \{n : |\omega_n - 1| \geq \varepsilon\} \cup \{0\}$ and the subspace $\mathfrak{M}_\varepsilon = \{\mu \in \mathfrak{M} : \mu\varphi_n = 0, n \notin \mathcal{N}_\varepsilon\}$. It is easily seen that $\mathfrak{M}_\varepsilon$ is a Banach space with the norm $\|\mu\|_\varepsilon^2 = \sum_{n \in \mathcal{N}_\varepsilon} |\mu\varphi_n|^2$ and the chain X satisfies condition (P) and is uniformly ergodic with respect to the norm $\|\cdot\|_\varepsilon$. But the sequence of operators P^{td} for any d does not converges in the operator norm as $t \to \infty$, since the spectrum of $P : \mathfrak{M}_\varepsilon \to \mathfrak{M}_\varepsilon$ contains the whole unit circle without ε-neighborhood of the unity that contradicts the condition (A4).

The proofs of the assertions listed above immediately follow from $P\varphi_n = \omega_n\varphi_n$, where $\{\varphi_n\}$ is an orthonormal system in the space $L_2(E)$ that is dual to $\mathfrak{M}$. Thus the matrix representation of $\overline{P}$ of the operator P coincides on $\mathfrak{M}$ with a diagonal matrix $\overline{P} = \operatorname{diag}(\omega_n, n \geq 0)$, $\overline{\Pi} = \operatorname{diag}(1, 0, 0, \ldots)$, the point spectrum of P coincides with the set $\{\omega_n\}$, and $\overline{P}^t - \overline{\Pi} = \operatorname{diag}(0, \omega_1^t, \omega_2^t, \ldots)$,

$$\overline{R} = \operatorname{diag}(0, (1 - \omega_1)^{-1}, \ldots).$$

Finally, removing columns and rows with indices $n \notin \mathcal{N}_\varepsilon$ we obtain the representations for restrictions of P, P^t, and R on $\mathfrak{M}_\varepsilon$.

COROLLARY 2.1. *Let condition* (C) *be satisfied. A chain X is aperiodic and uniformly ergodic with respect to the norm $\|\cdot\|$ if and only if one of the following equivalent conditions is valid*:

(B) $\|\mu P^n\| \leq \rho\|\mu\|$ *for all* $\mu \in \mathfrak{M} : \mu(E) = 0$ *with* $0 < \rho < 1$ *and some* $n \geq 1$;
(B1) $\|P^n - \Pi\| < 1$ *for some* $n \geq 1$;
(B2) $\|P^t - \Pi\| \to 0$ *as* $t \to \infty$;
(B3) $\|P^t - \Pi\| = O(\rho^t)$ *as* $t \to \infty$ *for some* $0 < \rho < 1$.

REMARK 2.3. Condition (B) for $\|\mu\| = |\mu|(E)$ is equivalent to the well known uniform mixing condition [Doob (1953), Davydov (1973), Revuz (1975)]:

$$\sup_{x,y \in E} \sup_{A \in \mathcal{E}} |P^n(x, A) - P^n(y, A)| \leq \rho < 1. \tag{2.6}$$

For $\|\mu\| = \|\mu\|_v$ (see 1.4), condition (B) is equivalent to the condition

$$\text{(Bv)} \qquad \int_E v(z)\, |P^n(x, dz) - P^n(y, dz)| \leq \rho(v(x) + v(y)) \tag{2.7}$$

for all $x, y \in E$ and for some $\rho < 1$.

To verify this equivalence note that measures of the form

$$\mu_{x,y}(A) = (\mathbb{I}_{x \in A} - \mathbb{I}_{y \in A}) / (v(x) + v(y))$$

are contained in the unit ball $\|\mu_{x,y}\|_v \leq 1$ and $\mu_{x,y}(E) = 0$. Thus (Bv) follows from (B), since any measure $\mu \in \{\mu \in \mathfrak{M}_v : \mu(E) = 0, \|\mu\| \leq 1\}$ can be represented as a convex hull $\mu(A) = \iint \mu_{x,y}(A)\, \Phi(dx, dy)$, where

$$\Phi(dx, dy) = (v(x) + v(y))\mu^+(dx)\mu^-(dy)/\mu^+(E)$$

and $\mu^{\pm}$ are the nonnegative orthogonal components in the Hahn decomposition of the measure μ with $\iint \Phi(dx,dy) = \|\mu^+\|_v + \|\mu^-\|_v = \|\mu\|_v \le 1$. Therefore condition (B) for component measures $\mu_{x,y}$ (that is an evident consequence of (Bv)) implies that (B) holds true for all μ with $\|\mu\|_v \le 1$, $\mu(E) = 0$.

REMARK 2.4. If the norm $\|\cdot\|$ additionally satisfies on $\mathfrak{M}$ the condition

(MC) there exists a set $C \in \mathcal{E}$ with $\pi(C) > 0$ such that the family of measures $\delta_x(A) = \mathbb{I}_{x\in A}$ belongs for $x \in C$ to $\mathfrak{M}$ and is bounded in the norm $\|\cdot\|$ and the σ-algebra $\mathcal{E}$ is separable

then the condition (C) in Theorem 2.2, Corollary 2.1 and in Theorems 2.3 and 2.4 can be omitted.

Condition (MC) holds true in particular for all weighted variation norms $\|\cdot\|_v$ introduced by (1.4).

Indeed, in view of Remark 2.3 it is sufficient to prove that (C) follows from condition (MC) and the uniform ergodicity of the chain X in the norm $\|\cdot\|$. Observe first that $\|\delta_x P^{(t)} - \pi\| \to 0$ uniformly on $x \in C$ as $t \to \infty$, since $\delta_x \Pi = \pi$, and $\sup(\|\delta_x\|, x \in C) < \infty$. Taking into account conditions (M3) and (M1) we find that $|\delta_x P^{(t)} - \pi|(E) \to 0$ uniformly on $x \in C$. Taking k to satisfy $|\delta_x P^{(k)} - \pi|(E) \le \pi(C)/2$, $x \in C$, we conclude that the measures $\delta_x P^{(k)}(A) = P^{(k)}(x,A)$ and $\pi(A)$ have a common component $\nu_x^{(k)}(A)$ with $\nu_x^{(k)}(E) \ge 1 - \pi(C)/2$ for all $x \in C$.

Denote by $\nu_x^t(A)$ the common component of the measures $P^t(x,A)$ and $\pi(A)$ and by $p_t(x,y)$ their density with respect to π. The relation $P^{t+1} = P^t P$ implies that $\nu_x^{t+1}(C) \ge \nu_x^t(C)$, since $\int \nu_x^t(dy)P(y,\cdot) \ll \pi(\cdot)$. Therefore

$$\nu_x^k(C) = \nu_x^k(E) - \nu_x^k(E\setminus C) \ge k^{-1}\sum_{t=0}^{k-1} \nu_x^t(E) - \pi(E\setminus C) = \nu_x^{(k)}(E) - 1 + \pi(C)$$
$$\ge 1 - \pi(C)/2 - 1 + \pi(C) = \pi(C)/2 > 0$$

for $x \in C$.

Put $\nu(A) = \pi(A \cap C)$. Then

$$\iint \nu(dx)\,\nu(dy) p_k(x,y) =\in t\nu(dx)\nu_x^k(C) \ge \pi^2(C)/2 > 0.$$

Hence, for some $\varepsilon > 0$, we get $\nu \times \nu(H) > 0$ where

$$H = \{(x,y) \in C \times C : p_k(x,y) \ge \varepsilon\}$$

and $\bigcup_{y\in C}\{x : (x,y) \in H\} = C$, $\varepsilon < \frac{1}{2}$. In the same way as in Revuz (1975), Lemma 6.1.1, we construct, for $F = G = H$, sets G_1, G_2, $G \in \mathcal{E} \cap C$ such that $\nu(G_i) > 0$, $\nu(G) > 0$, and $\nu\times\nu(H\cap(G_1\times G)) \ge 3\nu(G_1)\nu(G)/4$, $\nu\times\nu(H\cap(G\times G_2)) \ge 3\nu(G)\nu(G_2)/4$. In so doing, the lemma on differentiation of an integral and the separability of $\mathcal{E}$ should be used.

Put $H_x = \{y : (x,y) \in H\}$, $H^y = \{x : (x,y) \in H\}$ and define the sets

$$B_1 = \left\{x \in G_1 : \nu(G \cap H_x) \ge \frac{3}{4}\nu(G)\right\},$$
$$B_2 = \left\{y \in G_2 : \nu(G \cap H^y) \ge \frac{3}{4}\nu(G)\right\}.$$

Then $\nu(B_i) > 0$ because otherwise either

$$\nu \times \nu(H \cap (G_1 \times G)) = \int_{G_1} \nu(dx)\,\nu(G \cap H_x) < \frac{3}{4}\nu(G_1)\nu(G)$$

by the Fubini theorem or

$$\nu \times \nu(H \cap (G \times G_2)) < \frac{3}{4}\nu(G)\nu(G_2).$$

Let $x \in B_1$, $y \in B_2$. Definitions of H and p_k yield

$$\begin{aligned} p_{2k}(x,y) &\geq \int p_k(x,z)\,\pi(dz)p_k(z,y) \geq \varepsilon^2\nu(H_x \cap H^y) \geq \varepsilon^2\nu((H_x \cap G) \cap (G \cap H^y)) \\ &= \varepsilon^2\left(\nu(H_x \cap G) + \nu(G \cap H^y) - \nu((H_x \cap G) \cup (G \cap H^y))\right) \\ &\geq \varepsilon^2\left(\frac{3}{4}\nu(G) + \frac{3}{4}\nu(G) - \nu(G)\right) = \varepsilon^2\nu(G)/2 = \beta > 0. \end{aligned}$$

Putting $\alpha(A) = \nu(A \cap B_2)/\nu(B_2)$, $n = 2k$, $h(x) = \beta\nu(B_2)\mathbb{I}_{x \in B_1}$ we deduce that $P^n(x,A) \geq h(x)\alpha(A)$ for all $x \in E$, $A \in \mathcal{E}$, and $\alpha(E) = 1$, $\pi h = \beta\nu(B_2)\nu(B_1) > 0$. Note also that $P^{t+n} \geq P^t h \circ \alpha$ for all $t \geq 0$. If $\alpha P^t h = 0$ for all $t \geq n$ then $\alpha P^{(t)}h \to 0$ as $t \to \infty$. The measure α is concentrated on C and the measures $P^{(t)}(x,A)$ converge to the limit $\pi(A)$ uniformly in $x \in C$. Therefore $\alpha P^{(t)}h \to \alpha(E)\pi h > 0$ as $t \to \infty$. Thus $\alpha P^m h > 0$ for some $m \geq n$. Consequently, the inequality $P^{m+n} \geq h' \circ \alpha$ with $h' = P^m h$ implies condition (C).

Condition (MC) for norms $\|\cdot\|_v$ follows from the finiteness of the weight function v (it is sufficient to take $C_n = \{x: v(x) \leq n\}$. Then $\pi(C_n) > 0$ for some $n \geq 1$, since $C_n \uparrow E$ and $\|\delta_x\|_v = v(x) \leq n$ for $x \in C_n$.

PROOF OF THEOREM 2.2. First we show that conditions (A1)–(A2) and (A4)–(A6) are equivalent. To do this we prove the chain of implications

$$(A2) \Rightarrow (A1) \Rightarrow (A5) \Rightarrow (A4) \Rightarrow (A6) \Rightarrow (A2).$$

The first implication is evident. To prove the second one we observe that the space $\mathfrak{M}_d^0 = \{\mu \in \mathfrak{M}: \mu(E) = 0\} \cap \mathfrak{M}P^{(d)}$ is invariant with respect to the operator P. Put $\rho = \Lambda_n^d(P) < 1$. Then we obtain, by induction, for arbitrary $t \geq 1$ and for all $\mu \in \mathfrak{M}_d^0$ that

$$\left\|\mu P^{nt}\right\| = \left\|\mu P^{nt-n}P^n\right\| \leq \rho\left\|\mu P^{nt-n}\right\| \leq \cdots \leq \rho^t\|\mu\|.$$

It is easily seen that $\mu = \nu(I - \Pi)P^{(d)} \in \mathfrak{M}_d^0$ for all $\nu \in \mathfrak{M}$. So

$$\left\|\nu P^{(d)}\left(P^{tn} - \Pi\right)\right\| = \left\|\nu(I - \Pi)P^{(d)}P^{(nt)}\right\| \leq \rho^t\|\mu\| \leq \rho^t\|\nu\|\left\|P^{(d)} - \Pi\right\|.$$

Hence by the definition of the operator norm $\|P^{(d)}(P^{nt} - \Pi)\| \leq \rho^t\|P^{(d)} - \Pi\|$. Therefore for all natural $k = nt + s$, $0 \leq s < n$,

$$\left\|P^{(d)}(P^k - \Pi)\right\| = \left\|P^{(d)}(P^{nt} - \Pi)P^s\right\| \leq \rho^t\left\|P^{(d)} - \Pi\right\|\max_{s<n}\|P^s\| = O(\rho_n^k)$$

as $k \to \infty$ where $\rho_n = \rho^{1/n} < 1$. The implication $(A1) \Rightarrow (A5)$ is derived.

Suppose that condition (A5) is satisfied. Then the operator-valued function

$$R_z^d = \sum_{t \geq 0} z^{-t-1} P^{(d)} (P^t - \Pi)$$

is analytical in the strong sense, [see Dunford and Schvartz (1953), Chapter 7], in the domain $D = \{z: |z| > \rho\}$. Since $zR_z^d = P^{(d)}(I - \Pi) + PR_Z^d z$, $(z-P)R_z^d = R_z^d(z-P) = P^{(d)} - \Pi$. Therefore the operator $\left(P^{(d)} - \Pi\right)(z-P)^{-1}$ is bounded for $z \in D$ and coincides with R_z^d. Moreover, identities $(z-P)Q_z^d = Q_z^d(z-P) = z^{(d)} - P^{(d)}$, where $z^{(d)} = d^{-1}\sum_{s<d} z^s$, $Q_z^{(d)} = d^{-1}\sum_{s<d} z^s \sum_{l<d-s} P^l$ imply that $\left(z^{(d)} - P^{(d)}\right)(z-P)^{-1} = Q_z^d$ is an analytical function in D. Since $\Pi(z-P)^{-1} = \Pi(z-1)^{-1}$ for all $z \neq 1$, the resolvent of the operator P is represented in the form

$$\begin{aligned}(z-P)^{-1} &= \left(z^{(d)}\right)^{-1} \left((z-1)^{-1}\Pi + R_z^d + Q_z^d\right) \\ &= \left(z^d - 1\right)^{-1} \left(\Pi + (z-1)R_z^d + (z-1)Q_z^d\right).\end{aligned}$$

This expression evidently is a finite and meromorphic function on D and has only simple poles at the points $\{z: z^d = 1\}$. Therefore condition (A4) is fulfilled.

To prove the implication $(A4) \Rightarrow (A6)$ we denote by $\omega_1, \ldots, \omega_d$ the pole of the resolvent $P_z = (z-P)^{-1}$ of the transition operator P in the domain $D = \{z: |z| > \rho\}$ and by $\Pi_1, \ldots, \Pi_d$ the corresponding orthogonal projectors. By Theorem 7.3.18 from Dunford and Schvartz (1953) the operator function

$$Q_z = P_z - \sum_{j=1}^{d} (z - \omega_j)^{-1} \Pi_j$$

is analytical in D. Take $\beta \in (0,1)$ and use the Cauchy formula, [see Dunford and Schvartz (1953), Chapter 7], to obtain that for arbitrary $s \geq 1$

$$\begin{aligned}P^s &= \frac{1}{2\pi i} \int_{\beta|z|=1} z^s (z-P)^{-1}\, dz = \frac{1}{2\pi i} \int_{\beta|z|=1} \left(Q_z + \sum_{j=1}^{d} (z-\omega_j)^{-1}\Pi_j\right) z^s\, dz \\ &= \sum_{j=1}^{d} \omega_j^d \Pi_j + \frac{1}{2\pi i} \int_{|z|=\beta} z^s Q_z\, dz.\end{aligned} \tag{2.8}$$

Note that for $s = td$, $\omega_j^d = \left(\omega_j^d\right)^t = 1$, and then, for the projector $\Pi_0 = \sum \Pi_j$, it follows from (2.8) that

$$\left\|P^{td} - \Pi_0\right\| \leq \beta^s \sup_{|z|=\beta} \|Q_z\|, \tag{2.9}$$

where the right-hand side is finite, since Q_z is analytical in D and $\{z: |z| = \beta\} \subset D$.

It remains to check that the dimension of the projector Π_0 does not exceed d. Consider the operator mapping $P \to P^d$ and the space $\mathfrak{M}\Pi_0$ corresponding to the spectral set $\{\omega_1, \ldots, \omega_d\}$ of the operator P. According to Theorem 7.3.19 from Dunford and Schvartz (1953) on the spectrum transformation, the spectral space above coincides with the space $\{\mu \in \mathfrak{M}: \mu = \mu P^d = \omega^d \mu\}$. The dimension of the latter space does not exceed d, by Lemma 2.2. Therefore $\dim \mathfrak{M}\Pi_0 \leq d$ and the implication $(A4) \Rightarrow (A6)$ is proved.

Let condition (A6) be satisfied. Then $P^{td} \to \Pi_0$ as $t \to \infty$ and $\Pi_0 = \Pi_0^2 = \Pi_0 P^d = P^d \Pi_0$, where Π_0 commutes with P. Now we conclude that

$$0 = \Pi_0(I - P^d)d^{-1} = \Pi_0 P^{(d)}(I - P),$$

so $\Pi_0 P^{(d)}$ is an invariant orthogonal projector of the kernel P. Since the invariant measure is unique, $\Pi_0 P^{(d)} = (\Pi_0 P^{(d)} \mathbb{I}) \circ \pi = \Pi$ (see Remark 1.2). Note that for any measure $\mu \in \mathfrak{M}_d^0 = \{\mu \in \mathfrak{M}P^{(d)}: \mu(E) = 0\}$

$$\mu\Pi_0 = \nu P^{(d)}\Pi_0 = \nu\Pi_0 P^{(d)} = \nu\Pi = \nu P^{(d)}\Pi = \mu\Pi = 0.$$

Let $t = kd + s$, $0 \leq s < d$. Then

$$\mu P^t = \mu P^{kd} P^s = \mu(P^{kd} - \Pi_0)P^s.$$

Therefore

$$\|\mu P^t\| \leq \|\mu\| \max_{s<d} \|P^s\| \cdot \|P^{kd} - \Pi_0\|.$$

Taking into account the definition (2.5) we conclude that $\Lambda_t^d(P) \to 0$ as $t \to \infty$. So the condition (A2) is satisfied.

Now we prove that (A3) is equivalent to other conditions of the theorem. We check the implications $(A6) \Rightarrow (A3) \Rightarrow (A4)$. The first of these is evident (just take the finite-dimensional projector Π as the operator K).

Let (A3) hold true. Then the operator $\omega - P^n + K$ has a bounded inverse for $|\omega| = 1 > \|P^n - K\|$ and is a Fredholm operator. By Theorem 4.5.26 in Kato (1966) the compactness of the operator K implies that the operator $\omega - P^n$ is a Fredholm operator. According to Theorem 4.5.31 in Kato (1966) operators $z - \omega + \omega - P^n$, for all sufficiently small $|z - \omega| > 0$, have also the Fredholm property, and their defect does not depend on z. Since $(z - P^n)^{-1} \in \mathcal{B}$ for $|z| > 1$, this defect vanishes for small $|z - \omega| > 0$. Therefore all the points of the spectrum $\sigma(P^n)$ of the operator P^n that lie on the unit circle $\{z: |z| = 1\}$ are isolated and there exists $\beta \in (0,1)$ such that the set $\Omega_n = \sigma(P^n) \cap \{z: |z| > \beta\}$ is finite and $|\omega| = 1$ for $\omega \in \Omega_n$. Moreover, it follows from the Fredholm property of the operator $\omega - P^n$ that for $\omega \in \Omega_n$ the proper subspace of the operator P_n that is connected with the spectral set Ω_n has a finite dimension. Thus, by the spectrum transformation theorem, the set $\Omega = \sigma(P) \cap \{z: |z| > \rho\}$ with $\rho = \beta^{1/n} < 1$ is also finite and the corresponding projector Π_0 is finite-dimensional. The space $M = \mathfrak{M}\Pi_0$ is invariant for P and is generated by the cone $M^+ = M \cap \mathfrak{M}^+$. Choose an orthogonal basis $\{\alpha_1, \ldots, \alpha_l\}$ in M^+ and consider the representation $\alpha_i P = \sum \lambda_{ij}\alpha_j$. As in Lemma 2.3 we prove that the matrix $\Lambda = (\lambda_{ij})$ is stochastic and $\alpha_i P^s = \sum \lambda_{ij}^{(s)} \alpha_j$ where $(\lambda_{ij}^{(s)}) = \Lambda^s$. Therefore

$$\alpha_i(z - P)^{-1} = \sum z^{-t-1}\alpha_i P^t = \sum \lambda_{ij}(z)\alpha_j,$$

where $(\lambda_{ij}(z)) = \Lambda(z) = (z-\Lambda)^{-1}$. So the poles of the function $\Pi_0(z-P)^{-1}$ coincide with the poles of $\Lambda(z)$. Then, the second term in the representation $(z-P)^{-1} = \Pi_0(z-P)^{-1} + (I-\Pi)(z-P)^{-1}$ is analytical for $|z| > \rho$ by the definition of the projector Π_0. Hence the poles of the resolvent $(z-P)^{-1}$ coincide in the domain $\{z: |z| > \rho\}$ with the corresponding poles of the matrix function $\Lambda(z) = (z-\Lambda)^{-1}$, that is, they are the eigenvalues of the stochastic matrix Λ.

By the Frobenius theorem, [see Gantmakher (1967), Theorem 13.2.2], these values coincide with simple roots of unity and correspond to simple poles of the resolvent. So condition (A4) is valid.

It remains to prove the last assertion of the theorem. Note that (A4) is fulfilled for $d = d(X)$ and does not hold true for $d < d(X)$. Indeed, it follows from (2.1) that $\sigma(P) \cap \{z: |z| = 1\}$ contains those roots of the unity that are of an order less than $d(X)$. On the other hand, it is easily seen (as in the proof of Theorem 2.1) that $\omega \in \sigma(P)$, $\omega^d = 1$, implies $\omega^s \in \sigma(P)$ for $1 \le s \le d$. Therefore $\sigma(P) \cap \{z: |z| = 1\} = \{\omega^s, 1 \le s \le d(X)\}$ and all conditions of the theorem are satisfied for $d = d(X)$ and are not for $d < d(X)$ (because they are equivalent). The dimension of the projector $\Pi_0 = \Pi_0 P^d$ coincides with $d(X)$ by the definition.

To finish the proof of the theorem, note that for $t = kd + s$, $0 \le s < d$, the operator

$$P^{(t)} - \Pi = \frac{1}{k} \sum_{j=0}^{k-1} P^{(d)} \left(P^{jd} - \Pi_0\right) P^s \to 0$$

in operator norm as $t \to \infty$, provided condition (A5) is satisfied. Therefore (A1)–(A6) imply the uniform ergodicity of the chain X. Note that condition (C) was not used above.

It remains to prove that the uniform ergodicity of the chain X implies, for example, condition (A3). Provided condition (C) a stronger statement will be proved below in Theorem 2.4 with a finite-dimensional operator K in (A3). According to this theorem one can choose $K = P^{mnd} - T^m = (T+H)^m - T^m$, where the operator T and d-dimensional projector H are introduced above. The compactness of K is evidently derived from the finite-dimensionality of the projector H by removing the parenthesis in the product $(T+H)^m$. Observe that Theorem 2.2 and its corollaries are not used in the proof of the necessity of conditions of the theorem 2.4 (while constructing T and H).

Theorem 2.2 is proved. Corollary 2.1 is evident. □

2.4. Ergodicity criterion in terms of a residual kernel

By Theorem 1.3 the convergence in Definitions 1.1 and 1.4 is essentially connected with properties of the generalized potential R. Therefore one can expect that the rate of this convergence may be expressed in terms of the norm of the potential (R), (R3). However, the convergence in (B3) depends also on aperiodicity of the chain X (i.e., on the behaviour of the resolvent in a neighborhood of the whole unit circle). Thus the investigation of the rate of convergence in (B3) is necessarily based on numerical characteristics of the aperiodicity. On the other hand, we face some difficulties while checking (R)–(R3) for certain classes of Markov chains. For this reason we consider the following criterion that simplifies the analysis for many cases and gives an effective measure of the ergodicity.

THEOREM 2.3. *A Markov chain X that satisfies condition* (C) *is uniformly ergodic in the norm* $\|\cdot\|$ *and aperiodic if and only if for some* n, α, *and* h *from condition* (C)

(T) $\|T^m\| \leq \rho$ *for some* $m \geq 1$, $\rho < 1$, *where the kernel*

$$T = P^n - h \circ \alpha$$

is nonnegative according to (C).

Furthermore, the uniform ergodicity and aperiodicity of the chain X *imply that condition* (T) *is fulfilled for all* n, α, *and* h *satisfying condition* (C).

REMARK 2.5. Conditions (C) and (T) for $\|\mu\| = |\mu|(E)$ are equivalent to the well known Doeblin condition for the kernel P to be quasicompact, [see Neveu (1964), Chapter 6.3].

Condition (T) for the class of weighted variation norm $\|\cdot\|_v$ introduced in (1.4) is equivalent to the condition

(Tv) $T^m v(x) \leq \rho v(x)$ for all $x \in E$ and some $m \geq 1$, $\rho < 1$.

Therefore, the choice of a corresponding norm for a given chain is reduced to the construction of a test function $v(x)$.

PROOF OF THEOREM 2.3. The sufficiency of conditions (C) and (T) for the uniform ergodicity and aperiodicity immediately follows from Theorem 2.2. Indeed, let n, α and h satisfy (C). Then $h \circ \alpha \in \mathcal{B}$, $T \in \mathcal{B}$ by Lemma 1.1 and $\alpha \in \mathfrak{M}^+$, $h \in \mathfrak{N}^+$. Consider the operator

$$K = P^{nm} - T^m = (T + h \circ \alpha)^m - T^m = T^{m-1} h \circ \alpha + \cdots + h\alpha T^{m-1}.$$

It is clear that K is of a finite dimension, so it is compact. Under condition (T) we have $\|P^{nm} - K\| = \|T^m\| < 1$, that is, the chain X is uniformly ergodic by Theorem 2.2 (condition (A3)). It remains to check that the chain X is aperiodic. This immediately follows from the aperiodicity of a chain with the transition kernel P^n, since the period of this chain does not depend on n. Therefore it can be assumed that $n = 1$. Let a chain with the kernel P have a period $d > 1$. Then by Lemma 2.3 there exists a sequence of pairwise orthogonal measures π_i such that $\pi_i P = \pi_{i+1}$ for $1 \leq i \leq d$ and $\pi_d P = \pi_1$. Hence $\pi_{i+1} \geq \pi_i(h \circ \alpha) = (\pi_i h)\alpha$ and $\pi_{i+1} h \geq \varepsilon \pi_i h$, where $\varepsilon = \alpha h > 0$ by (C). Relations $\pi h > 0$ and (2.3) imply $\pi_i h > 0$ for at least one i. Thus $\pi_i h > 0$ for all i. Now the inequalities $\pi_{i+1} \geq (\pi_i h)\alpha$ ensure that the measures π_i have a common nonzero component (that is proportional to α) and cannot be orthogonal. This contradiction shows that $d = 1$.

The proof of the sufficiency of conditions (C) and (T) is based on the following assertion.

LEMMA 2.5. *Let* $Q \in \mathcal{B}$ *be a substochastic kernel and the operator* $I - Q$ *have a bounded inverse. Then the resolvent* $(z - Q)^{-1}$ *is analytical in a neighborhood of the unit circle and* $\|Q^m\| = o(\beta^m)$ *as* $m \to \infty$ *for some* $\beta < 1$.

PROOF. By the stability of the bounded invertibility it follows that $(I - uQ)^{-1} \uparrow (I - Q)^{-1}$ as $u \uparrow 1$. Since the operator $(I - uQ)^{-1} = \sum u^t Q^t$ is nonnegative, $(I - Q)^{-1} \in \mathcal{B}_+$. Let $\mu \in \mathfrak{M}^+$. Then

$$\sum_{t<n} \mu Q^t = \mu(I - Q)^{-1}(I - Q^n) \leq \mu(I - Q)^{-1} \in \mathfrak{M}^+$$

for all n. By condition (M1) we have $\mu(I-Q)^{-1}(E) < \infty$, so the measure $\sum \mu Q^t$ is finite and for all $A \in \mathcal{E}$

$$\sum \mu Q^t(A) = \lim_{n\to\infty} \sum_{t<n} \mu Q^t(A) \le \mu(I-Q)^{-1}(A).$$

Therefore $\sum \mu Q^t \in \mathfrak{M}^+$ for all $\mu \in \mathfrak{M}^+$ by condition (M2).

Consider for $\mu \in \mathfrak{M}$, $|z| = 1$, the sum of a series $\nu = \sum z^{-t-1}\mu Q^t$. It is evident that $|\nu| \le \sum |\mu Q^t| \le \sum |\mu| Q^t \in \mathfrak{M}^+$, since $|\mu| \in \mathfrak{M}^+$ by condition (M3). Then $|\nu| \in \mathfrak{M}^+$ according to (M2) and $\nu \in \mathfrak{M}$. The measure ν satisfies the equation $z\nu - \nu Q = \mu$. Therefore this equation, for all $|z| = 1$ and $\mu \in \mathfrak{M}$, has a solution $\nu \in \mathfrak{M}$. This solution is unique because otherwise from $z\nu = \nu Q$ follows $|\nu| \le |\nu| Q$ and $|\nu|(I-Q) \le 0$. So, finally, $|\nu| \le 0$, since $(I-Q)^{-1} \in \mathcal{B}^+$. Then the resolvent $\mu(z-Q)^{-1} \in \mathfrak{M}$ is well defined for all $\mu \in \mathfrak{M}$ and by the closed graph theorem, [see Rudin (1973), Theorem 2.15], we have $(z-Q)^{-1} \in \mathcal{B}$ for all $|z| = 1$. It does not intersect a neighborhood of the unit circle because the spectrum is closed.

The second assertion of the lemma follows from the Cauchy formula for an analytical function $(z-Q)^{-1}$ in the domain $|z| > \beta$ with $\beta < 1$. Details are the same as in proof of (2.9). The lemma is proved. □

Assume that the chain X is uniformly ergodic and aperiodic, and n, α, h satisfy condition (C). By Lemma 2.4 the operator $I - P^n$ has the Fredholm property. Then the operator $I - T = I - P^n + h \circ \alpha$ is also Fredholm by Theorem 4.5.26 in Kato (1966). Moreover, this operator has no annihilators. Indeed, let $\mu\mathbb{1} = 1$ and $\mu = \mu T$. Then $\mu\mathbb{1} = \mu P^n 1 - (\mu h)\alpha\mathbb{1} = \mu\mathbb{1} - \mu h$, that is, $\mu h = 0$. Hence $\mu = \mu T = \mu P^n$. Since the chain X is aperiodic, $\mu = c\pi$. On the other hand $0 < \pi h$, $\mu h = 0$. This contradiction shows that $I - T$ is an operator with the Fredholm property and with zero defect. Thus $(I-T)^{-1} \in \mathcal{B}$. It is clear that T is substochastic in view of (C). So $\|T^m\| \to 0$ by Lemma 2.5 and $\|T^m\| < 1$ for some m. The theorem is proved. □

The general periodic case is considered in the following theorem.

THEOREM 2.4. *Let condition* (C) *hold true. A chain X is uniformly ergodic in a norm $\|\cdot\|$ if and only if there exist a natural d, the measures $\alpha_1, \dots, \alpha_d \in m\mathcal{E}^+$, and functions $h_1, \dots, h_d \in f\mathcal{E}^+$ such that*

(Cd) $\pi h_i > 0$, $\alpha_i(E) = 1$, $\alpha_i h_i > 0$, *and the kernel* $T = P^{nd} = \sum_{i=1}^d h_i \circ \alpha_i$ *is nonnegative for some* $n \ge 1$;

(T) $\|T^m\| \le \rho$ *for some* $m \ge 1$, $\rho < 1$.

PROOF. The sufficiency of conditions (Cd) and (T) for the uniform ergodicity of the chain X can be proved in the same way as an analogous assertion in Theorem 2.4 (the operator $K = P^{nmd} - T^m$ is compact and satisfies condition (A3), since $H = \sum h_i \circ \alpha_i$ is compact).

Let us prove the necessity of this. Denote by $d = d(X)$ the period of the chain X and consider the sets E_i and the measures π_i that are constructed in Lemma 2.3. Let n, α, and h satisfy condition (C). Since $(\pi h)\alpha \le \pi P^n = \pi$ and $\pi(E_0) = 0$, $\alpha(E_0) = 0$, $\alpha h = \sum \int_{E_i} h(x)\,\alpha(dx) > 0$. Therefore at least one term of this sum is positive and without loss of generality one can assume that α is concentrated on E_1, $\alpha h > 0$, $\alpha(E_1) = 1$.

Note that condition (C) implies $P^{s+n+t} \geq P^s h \circ \alpha P^t$. Therefore

$$P^{nd} \geq d^{-1} \sum P^{nd-i+1} \geq P^s h \circ \alpha P^t,$$

where $\alpha_i = \alpha P^{i-1}$, $h_i = d^{-1} P^{nd-i+1} h$. It is easily seen that $\pi h_i = d^{-1} \pi h > 0$, $\alpha_i h_i = d^{-1} \alpha P^{nd} h \geq d^{-1} \alpha (h \circ \alpha)^d h = d^{-1} (\alpha h)^{d+1}$, and $\alpha_i(E) = 1$. Since the support of α is E_1, by Lemma 2.3 we get $\alpha_i(E_i) = 1$ for $1 \leq i \leq d$. Moreover, $\alpha_i \leq c\pi_i$, since $\alpha_1 \leq (\pi h)^{-1} \pi P^{i-1} = (\pi h)^{-1} \pi$, the support of α_i is E_i, and equality (2.3) holds true.

As in Theorem 2.3, from Lemma 2.4 we conclude that the operator $I - T = I - P^n + \sum h_i \circ \alpha_i$ possesses the Fredholm property. But it has no invariant measures. Indeed, let $\mu = \mu T$. Then by Lemma 2.1 one can assume that $\mu \geq 0$. Let μ_i be the restriction of the measure μ on the set E_i. Then $\mu = \mu P^{nd} - \sum (\mu h_i) \alpha_i$ and $\mu_i \leq \mu_i P^{nd} - (\mu h_i) \alpha_1$, since the support of α_j and $\mu_j P^{nd}$ is E_j by Lemma 2.2. Therefore $\mu_i \mathbb{I} \leq \mu_i P^{nd} \mathbb{I} - (\mu h_i) \alpha_1 \mathbb{I} = \mu_i \mathbb{I} - \mu h_i$. Hence $\mu h_i = 0$ for all i and $\mu = \mu P^{nd}$. By the definition of the period and by the construction of measures π_i given in Lemma 2.2 $\mu = \sum c_i \pi_i$ with $c_i \geq 0$. So $\mu h_i \geq c_i(\pi_i h_i) \geq c^{-1} c_i (\alpha_i h_i)$. Since $\alpha_i h_i > 0$ and $\mu h_i = 0$, $c_i = 0$ for all i. Therefore $\mu = 0$ and the kernel $I - T$ has no nontrivial annihilators. Thus $(I-T)^{-1} \in \mathcal{B}$ and Lemma 2.4 implies condition (T). The theorem is proved. □

COROLLARY 2.2. *Let $E = \{0, 1, \ldots\}$, the transitions of a chain X be given by the matrix $P = (p_{ij}, i, j \geq 0)$, and the norm of a measure $\mu = (\mu_j)$ be equal to $\|\mu\|_v = \sum |\mu_j| v_j$, where the sequence (v_j) is bounded from below: $\inf v_j > 0$.*

The chain X is uniformly ergodic in the norm $\|\cdot\|_v$ if and only if there exists some $m \geq 1$ such that

$$\limsup_{i \to \infty} \frac{1}{v_i} \sum_{j \in E} p_{ij}^m v_j < 1, \tag{2.10}$$

where p_{ij}^m are the transition probabilities over m steps.

The proof of the corollary immediately follows from Theorem 2.4 with α_i to be replaced by $(p_{ij}^m, j \in E)$, $h_i = (\delta_{ij}, j \in E)$, and the operator norm $\|\cdot\|_v$ is evaluated by (1.5).

COROLLARY 2.3. *Let conditions of the previous corollary be satisfied with $v_j = v^j$, where $v \geq 1$ is a numeric parameter.*

Then a chain X is uniformly ergodic in a norm $\|\cdot\|_v$ if there exists some $u > 1$ such that

$$\limsup_{i \to \infty} \sum (j - i) p_{ij} < 0, \tag{2.11}$$

$$\sup_{i \geq 1} \sum u^{j-i} p_{ij} < \infty, \tag{2.12}$$

Indeed, (2.12) implies $\|P\|_v < \infty$ for all $v \in [1, u]$. Substituting $v_j = v^j$ and $m = 1$ in (2.10) we obtain

$$\sum p_{ij} v^{j-i} = 1 + \ln v \sum (j - i) p_{ij} + O(\ln^2 v)$$

uniformly on i as $v \to 1$, since the function $(v^i - 1 - i \ln v)/\ln^2 v$ is dominated by u^i for $v \leq u$ and condition (2.12) holds true. Therefore the upper limit for sufficiently small $\ln v$ in (2.10) is strictly less than unity. So the uniform ergodicity of the chain X is derived.

CHAPTER III

UNIFORM ESTIMATES OF CONVERGENCE IN ERGODIC AND STABILITY THEOREMS

The conditions for the uniform ergodicity and the strong stability of a Markov chain introduced above are expressed in terms of "ergodic" functionals of the transition kernel such as a norm of the generalized potential, a uniform mixing coefficient, etc. It was shown that the boundedness of these functionals is equivalent to the property of the uniform ergodicity. Moreover, the boundedness of ergodicity functionals defined in the previous Chapter implies the uniformity of the ergodicity not only with respect to the initial distribution but also with respect to corresponding classes of Markov chains.

We obtain below the uniform estimates of convergence in ergodic theorems for wide classes of Markov chains. These estimates are based on criterion (R) of Theorem 1.3 (for the boundedness of a norm of the potential operator), on criterion (B) of Corollary 2.1 (that the ergodicity coefficient is less than unity), and on criteria (T) of Theorem 2.3 (that a norm of the residual kernel is less than unity).

We assume in this Chapter that a Markov chain $X = (X_t, t \geq 0)$ on $(E, \mathcal{E})$ given by its transition kernel $P(x, A)$ has a unique invariant stochastic measure π and satisfies condition (P), so its transition operator is bounded.

Along with X, we also consider a Markov chain $Y = (Y_t, t \geq 0)$ with the transition kernel $Q(x, A)$. Let ν denote a unique invariant stochastic measure for Q. The uniqueness of this measure follows from Theorem 1.5 if the chain Y is uniformly ergodic and the operator norm of the perturbation kernel $D(x, A) = Q(x, A) - P(x, A)$ is sufficiently small.

3.1. Estimates in terms of the potential operator

According to Theorem 1.3, the uniform ergodicity of the chain X implies that the stationary projector Π belongs to $\mathcal{B}$ and the generalized potential R belongs to $\mathcal{B}$, too. Let us introduce the following numerical ergodicity indices

$$\sigma = ||\Pi||, \qquad r = ||R||, \tag{3.1}$$

where the operator norm is defined by (1.3). Note that criteria (R3) of Theorem 1.3 gives us an explicit estimate for (3.1).

It is clear that the finiteness of (3.1) does not imply the convergence of powers P^t in the periodic case. So, in this paragraph we consider only the Cesaro averages $P^{(t)} = t^{-1} \sum_{s<t} P^s$.

To state the main result, we define, for $u \in (0, 1)$, the real function $g(u) = 1/u$ for $u \geq e^{-1}$ and $g(u) = 0$ otherwise. Let g_1 and g_2 be the polynomials of a degree that does not exceed m and such that $g_1(u) \leq g(u) \leq g_2(u)$ for all $u \in (0, 1)$. Denote also by $\varepsilon = \int_{[0,1]} (g_2(u) - g_1(u))\, du$ the approximation accuracy in $L_1[0, 1]$ of the function g by means of polynomials g_1, g_2. Let $G = |g_1(1)| + g_2(1)$ and S be the sum of absolute values of coefficients of the polynomials g_1, g_2, and their first derivatives.

THEOREM 3.1. *Let conditions* (M1)–(M3) *be satisfied and the chain be uniformly ergodic in the norm* $\|\cdot\|$, *so that the indices* (3.1) *are finite. Then for all* $t > t_0 = (m+1)(r+1)$

$$\left\|P^{(t)} - \Pi\right\| \le 2\sigma\varepsilon + \frac{2r+\sigma}{t}G + \left(\frac{\sigma}{2t^2} + \frac{2\sigma^2}{t(t-t_0)}\right)S. \tag{3.2}$$

COROLLARY 3.1. *Let the norm* $\|\cdot\|$ *on* $\mathfrak{M}$ *satisfy conditions* (M1)–(M3). *Then* $\left\|P^{(t)} - \Pi\right\| = O\left((\ln t)^{-1/4}\right)$ *as* $t \to \infty$ *uniformly on any class of Markov chains with bounded norms of the kernel* P, *of the projector* Π, *and of the potential* R.

PROOF OF COROLLARY 3.1. Denote by

$$B_m(g,u) = \sum_{k=0}^{m} \binom{m}{k} (1-u)^{m-k} u^k g\left(\frac{k}{m}\right)$$

the Bernstein polynomials. Define the continuous functions h_1 and h_2 as follows: $h_1(u) = g(u) + \varepsilon$ for $u \notin \left(e^{-1} - \varepsilon, e^{-1}\right)$ and $h_2(u) = g(u) - \varepsilon$ for $u \notin \left(e^{-1}, e^{-1} + \varepsilon\right)$. We linearly extend these functions to the corresponding ε-neighborhood of the point e^{-1}. We also put $g_i(u) = B_m(h_i, u)$, $i = 1, 2$. It is well known that

$$|h(u) - B_m(h,u)| \le c\left(\sup \frac{1}{m}|h'(u)|^2\right)^{1/2}$$

for all piece-wise differentiable functions h. Since $\sup|h_i'(u)| \sim \varepsilon^{-1}$, $\sup|g_i(u) - h_i(u)| \le \varepsilon$ for $m \sim \varepsilon^{-4}$. By the definition of h_i we conclude that $g_1 \le g \le g_2$, $\int_{[0,1]}(g_2 - g_1)\,du = O(\varepsilon)$, and $G = |g_1(1)| + g_2(1)$ is uniformly bounded, and thus the sum $S = O\left(m2^m\right) = O\left(\varepsilon^{-4}e^{c\varepsilon^{-4}}\right)$. We note that only two constants σ and r on the right-hand side of (3.2) depend on the chain X and on the norm $\|\cdot\|$. By the assumption, they are uniformly bounded. Substituting in (3.2) $\varepsilon = \beta(\ln t)^{-1/4}$, we immediately obtain that the right-hand side of (3.2) is $O\left((\ln t)^{-1/4}\right)$ for a sufficiently small β. □

COROLLARY 3.2. *Under conditions of Theorem* 3.1, *there exist absolute constants* a *and* b *such that*

$$\sup_{t \ge 0} \frac{1}{1+t}\left\|P^t\right\| \le b\left(r + \sigma^2 + \|P\|^{ar}\right).$$

To prove Corollary 3.2 it is sufficient to observe that

$$P^t = t\left(P^{(t)} - \Pi\right) - (t-1)\left(P^{(t-1)} - \Pi\right) - \Pi$$

and apply (3.2).

PROOF OF THEOREM 3.1. The proof of Theorem 3.1 is a modification of an Karamata idea, [see Titchmarsh (1952), Theorem 7.5.3], for the case of positive operators instead of positive reals.

Let $u \in (0,1)$. Then by Definition 1.3 the following identities are valid: $R(I-P) = I - \Pi$, $(u+(1-u)R)(I-\Pi) = R(I-uP)$, $R\Pi = 0$. Hence, for $r(1+r)^{-1} < u < 1$, we get

$$\begin{aligned}(I-uP)^{-1} &= \frac{1}{1-u}\Pi + (u+(1-u)R)^{-1}R \\ &= \frac{1}{1-u}\Pi + \frac{1}{u}R - (1-u)\frac{1}{u}R^2(u+(1-u)R)^{-1},\end{aligned} \tag{3.3}$$

where the operator $u+(1-u)R$ has a bounded inverse, since $u^{-1}(1-u)\|R\| < 1$.

For any bounded function φ on $[0,1]$ and for all $\theta \in (0,1)$ consider the operator on $\mathfrak{M}$ of the form

$$R_\theta(\varphi) = \sum_{n\geq 0} \theta^{n+1}\varphi\left(\theta^{n+1}\right)P^n. \tag{3.4}$$

The series in (3.4) converges by Lemma 1.2, so $R_\theta(\varphi) \in \mathcal{B}$. Moreover, $R_\theta(\varphi)$ linearly depends on φ and $R_\theta(\varphi) \in \mathcal{B}^+$ provided the function φ is nonnegative. Note that $R_\theta(g) = tP^{(t)}$ if $\theta = \exp\{-1/t\}$ for the function g defined as above. For the corresponding polynomials g_1 and g_2 we have $R_\theta(g) - R_\theta(g_1) = R_\theta(g-g_1) \geq 0$ and $R_\theta(g_2 - g) \geq 0$. Therefore, for $\theta = \exp\{-1/t\}$,

$$R_\theta(g_1) - t\Pi \leq t\left(P^{(t)} - \Pi\right) \leq R_\theta(g_2) - t\Pi.$$

Lemma 1.1 yields

$$t\left\|P^{(t)} - \Pi\right\| \leq 2\|R_\theta(g_1) - t\Pi\| + 2\|R_\theta(g_2) - t\Pi\|. \tag{3.5}$$

It remains to estimate the right-hand side of (3.5). Note that, for the polynomial $g_1(u) = \sum_{i=0}^m g_{1i}u^i$, the following identities are valid

$$\begin{aligned}R_\theta(g_1) &= \sum_{i=0}^m g_{1i}\sum_{n\geq 0}\theta^{n+1}\theta^{ni+i}P^n = \sum_{i=0}^m g_{1i}\theta^{i+1}(I-\theta^{i+1}P)^{-1} \\ &= \sum_{i=0}^m g_{1i}\left(\theta^{i+1}(1-\theta^{i+1})^{-1}\Pi + R\right) \\ &\quad - \sum_{i=0}^m g_{1i}R^2\left(\theta^{i+1} + (1-\theta^{i+1})R\right)^{-1}(1-\theta^{i+1}),\end{aligned} \tag{3.6}$$

where the latter equality follows from (3.3). In (3.6)

$$1 > \theta^{i+1} = \exp\left\{-\frac{i+1}{t}\right\} \geq 1 - \frac{i+1}{t} \geq 1 - \frac{m+1}{t} > \frac{r}{1+r} \tag{3.7}$$

provided that $t > t_0 = (m+1)(r+1)$. Transforming (3.6), one can write down the following representation

$$\begin{aligned}R_\theta(g_1) - t\Pi &= \Pi\sum_{i=0}^m g_{1i}\left(\theta^{i+1}(1-\theta^{i+1})^{-1} - \frac{t}{i+1} + \frac{1}{2}\right) \\ &\quad + \left(R - \frac{1}{2}\Pi\right)g_1(1) + t\Pi\left(\sum_{i=0}^m g_{1i}\frac{1}{1+i} - 1\right) \\ &\quad - R^2\sum_{i=0}^m g_{1i}(1-\theta^{i+1})\left(\theta^{i+1} + (1-\theta^{i+1})R\right)^{-1}.\end{aligned} \tag{3.8}$$

Note that $\theta = \exp\{-1/t\}$ and the coefficient for g_{1i} of the first summand in (3.8) equal to $a(u) = \exp\{-u\}/(1 - \exp\{-u\} - u^{-1} + 1/2$ for $u = (i+1)/t$, $u \in (0, 1)$, and $t > t_0$. One can prove that $0 \le a(u) \le u/6$ for all $u \in (0, 1)$. Furthermore, the expression in the parenthesis in the third summand in (3.8) is equal to

$$\sum_{i=0}^{m} \frac{g_{1i}}{i+1} - 1 = \int_{[0,1]} g_1(u)\, du - 1 = -\int_{[0,1]} (g - g_1)\, du.$$

Finally, for $u \ge 1 - (i+1)/t$, $i \le m$, and $t > t_0$ (in particular, $u > r(1+r)^{-1}$)

$$\begin{aligned} \|(1-u)(u + (1-u)R)^{-1}\| &\le \frac{1-u}{u} \frac{1}{1 - \frac{1-u}{u} r} = \frac{1-u}{u - (1-u)r} \le \frac{i+1}{t - (i+1)(r+1)} \\ &= \frac{1-u}{t - (i+1)(r+1)} \le \frac{i+1}{t - t_0}. \end{aligned} \tag{3.9}$$

Using the above remarks and the identities (3.9), (3.7), we obtain from (3.8) that for all $t > t_0$

$$\begin{aligned} \|R_\theta(g_1) - t\Pi\| &\le \sigma \sum_{i=0}^{m} |g_{1i}| \frac{1+i}{6t} + \left(r + \frac{\sigma}{2}\right) |g_1(1)| \\ &+ \sigma t \int_{[0,1]} (g(u) - g_1(u))\, du + r^2 \sum_{i=0}^{m} |g_{1i}| \frac{1+i}{t - t_0}. \end{aligned} \tag{3.10}$$

A similar inequality can also be proved for $\|R_\theta(g_2) - t\Pi\|$. Summing up these estimates and substituting the result into (3.5) we obtain (3.2). The theorem is proved. □

Using a bit more complicated functional instead of r one can obtain a slightly better estimate compared with (3.2).

THEOREM 3.2. *Let condition* (M1) *be satisfied. Assume that* $r = \|R\| < \infty$ *and* $p = \sup_{t \ge 0} \|P^t\| < \infty$. *Then for all* $t \ge 1$

$$\left\| P^{(t)} - \Pi \right\| \le \frac{r(1+p)}{t}. \tag{3.11}$$

REMARK 3.1. The functional p is finite, in particular, if the chain X is uniformly ergodic, condition (C) (or (MC)) is satisfied, and the norm $\|\cdot\|$ satisfies conditions (M1)–(M3). This statement immediately follows from Theorem 2.2, item (A6). Furthermore, $p = 1$ for the total variation norm $\|\mu\| = |\mu|(E)$.

PROOF. In view of (6.7), $(P^{(t)} - \Pi)(I - P) = t^{-1}(I - P^t)(I - \Pi)$, $(P^{(t)} - \Pi)\Pi = 0$. Then the measure $\mu(P^{(t)} - \Pi)$ satisfies Definition 1.3 with the measure

$$\nu = t^{-1}\mu(I - P^t)$$

on the right-hand side. Thus

$$P^{(t)} - \Pi = \frac{1}{t} R(I - P^t). \tag{3.12}$$

Evaluating the norm in (3.12) and estimating $\|I - P^t\| \le 1 + p$ we obtain (3.11). □

THEOREM 3.3. *Let σ and r in* (3.1) *be finite. Then any stochastic kernel $Q = P + D$ from a neighborhood $\{Q\colon \|D\| = \varepsilon < r^{-1}\}$ has a unique invariant stochastic measure $\nu \in \mathfrak{M}$ such that*

$$\begin{aligned} \|\nu - \pi\| &\le \varepsilon r\|\nu\| \le \varepsilon r \frac{\|\pi\|}{1-\varepsilon r}, \\ \|\nu - \pi - \pi D R\| &\le \varepsilon^2 r^2 \frac{\|\pi\|}{1-\varepsilon r}. \end{aligned} \tag{3.13}$$

Estimates that are more general than (3.13) will be obtained in theorems 5.1-5.3 below. Observe only that the conditions of Theorem 3.3 are fulfilled for all strongly stable Markov chains provided conditions (M1)–(M3) are satisfied. This statement follows from Theorems 1.3 and 1.5.

3.2. Estimates in terms of ergodicity coefficient

We assume in this paragraph that the following condition is valid

(B) $\|\mu P^n\| \le \rho\|\mu\|$ for all $\mu \in \mathfrak{M}$, $\mu(E) = 0$, and for some $n \ge 1$, $\rho < 1$.

By Corollary 2.1, condition (B) as well as the inclusion $\Pi \in \mathcal{B}$ is fulfilled for all uniformly ergodic aperiodic Markov chains.

THEOREM 3.4. *Let condition* (B) *be satisfied. Then for all $t \ge 0$*

$$\|P^t - \Pi\| \le p_n \rho^{[t/n]}, \tag{3.14}$$

$$\|R\| \le \frac{p'_n}{1-\rho}, \tag{3.15}$$

where $p_n = \max_{0\le s\le n} \|P^s - \Pi\| \le \sigma + \|P\|^{n-1}$, $p'_n = n\|P^{(n)} - \Pi\|$, and R is the generalized potential of the chain X.

PROOF. Denote by

$$\mathfrak{M}^0 = \{\mu \in \mathfrak{M}\colon \mu(E) = 0\}$$

the subspace of zero mass measures. The space $\mathfrak{M}^0$ is invariant for the stochastic kernel P. Hence (B) yields for $t = nk + s$, $0 \le s < n$,

$$\|\mu P^t\| = \|\mu P^s P^{nk-n} P^n\| \le \rho\|\mu P^s P^{nk-k}\| \le \cdots \le \rho^k \|\mu P^s\|.$$

It is easily seen that $\mu = \nu(I - \Pi) \in \mathfrak{M}^0$ for all $\nu \in \mathfrak{M}$. Then

$$\begin{aligned} \left\|\nu\left(P^t - \Pi\right)\right\| &= \|\nu(I-\Pi)P^t\| \le \rho^k\|\nu(I-\Pi)P^s\| = \rho^{[t/n]}\|\nu\left(P^s - \Pi\right)\| \\ &\le \|\nu\|\rho^{[t/n]} p_n, \end{aligned}$$

whence (3.14) follows because the right hand side is uniform by ν. The estimate for p_n follows from the definition of σ and from the inequality $\|P^s\| \le \|P\|^s \le \|P\|^{n-1}$, where $\|P\| \ge 1$ because of $P\mathbb{I} = 1$, $\mathbb{I} \in \mathfrak{N}$.

The inequality (3.15) is obtained from Definition 1.3. Let $\mu = \nu R$, so that $\mu(I - P) = \nu(I - \Pi)$ and $\mu\Pi = 0$. Then $\mu\mathbb{I} = 0$ and $\mu \in \mathfrak{M}^0$. Multiplying this equality by $nP^{(n)}$ we conclude that $\mu = \mu P^n + n\nu\left(P^{(n)} - \Pi\right)$. Therefore, $\|\mu\| \le \rho\|\mu\| + p'_n\|\nu\|$, $\|\mu\| \le \|\nu\|p'_n/(1-\rho)$. The estimate (3.15) is proved. □

THEOREM 3.5. *Let condition* (B) *be satisfied. Then any stochastic kernel Q from a neighborhood $\|Q^n - P^n\| = \varepsilon_n < 1 - \rho$ has a unique invariant stochastic measure ν, $\nu \in \mathfrak{M}$, that corresponds to a uniformly ergodic aperiodic Markov chain and such that*

$$\|\nu - \pi\| \leq \varepsilon_n \frac{\|\pi\|}{1 - \rho - \varepsilon_n}, \tag{3.16}$$

$$\|\nu - \pi - \pi(Q^n - P^n)R_n\| \leq \varepsilon^2 \|\pi\| \frac{1 + \sigma}{(1 - \rho)(1 - \rho - \varepsilon_n)}, \tag{3.17}$$

$$\sup_{t \geq 0} \left\|Q^t - P^t\right\| \leq \varepsilon_n \sigma \frac{q_n}{1 - \rho - \varepsilon_n} + \bar{\varepsilon}_n, \tag{3.18}$$

where

$$q_n = \max_{s<n} \|Q^s\|, \qquad \bar{\varepsilon}_n = \max_{s<n} \|Q^s - P^s\|,$$

R_n is the generalized potential of a chain with the kernel P^n.

REMARK 3.2. In view of the estimates

$$\varepsilon_n \leq \bar{\varepsilon}_{n+1} \leq (\|P\| + \varepsilon)^n - \|P\|^n, \tag{3.19}$$

it immediately follows from (3.16)–(3.18) that the inequalities are valid in terms of the deviation $\varepsilon = \|Q - P\|$. The condition $\varepsilon_n < 1 - \rho$ is essential for a pure stability effect of a invariant distribution. This is showed in the following example.

EXAMPLE 3.1. Let a chain $X_t = X_0 + \xi_1 + \cdots + \xi_t \mod 1$ with values in $E = [0, 1]$ be constructed in the same manner as in Example 2.1. Assume that a probability distribution $F(B) = \mathsf{P}(\xi_t \in B)$ has an absolutely continuous component, its density is not less than a on the set $C \in \mathcal{B}[0, 1]$, and $c = L(C) > \frac{1}{2}$. We take $\mathfrak{M}$ as the space of finite measures with the total variation norm $\|\mu\| = |\mu|(E)$. We prove below that the chain X satisfies condition (B) for $n = 1$ with the index $\rho = 1 - a(2c - 1) < 1$.

From Theorems 3.4 and 3.5 we derive for all $t \geq 1$, $B \in \mathcal{B}[0, 1]$, $x \in E$, that

$$|\mathsf{P}_x(X_t \in B) - L(B)| \leq (1 - ac)(1 - a(2c - 1))^{t-1}. \tag{3.20}$$

Any Markov chain Y with the transition kernel Q belonging to a neighborhood

$$\sup_{x \in E, B \in \mathcal{B}[0,1]} |Q(x, x + B) - F(B)| = \varepsilon < a\left(c - \frac{1}{2}\right) \tag{3.21}$$

is uniformly ergodic and meets the following uniform with respect to $1 \leq t \leq \infty$, $B \in \mathcal{B}[0, 1]$, $x \in E$, estimate

$$|\mathsf{P}_x(Y_t \in B) - \mathsf{P}_x(X_t \in B)| \leq \frac{\varepsilon}{a(2c - 1) - 2\varepsilon}. \tag{3.22}$$

Moreover, the supremum over B of the left-hand side of (3.20), for the distribution $F(B) = (1 - ac)\delta_{1/2}(B) + aL(B \cap [(1 - c)/2, (1 + c)/c])$, is not less than

$$\left(1 - ac\frac{q}{\pi} \sin \pi c\right)^t \geq \left(1 - ac + \frac{2a}{\pi}(1 - c)\right)^t, \tag{3.23}$$

where $\delta_x(\cdot)$ is the unit mass measure at the point x.

REMARK 3.2. For $c = 1$ the right-hand sides of (3.20) and (3.23) coincide with $(1-a)^t$, so the estimate (3.20) is precise in this case. Note also that for $c = 1$ the kernel Q may be chosen to have no invariant measures on the boundary of a neighborhood $\{Q: \varepsilon < \frac{1}{2}\}$.

The inequalities (3.20) and (3.22) immediately follow from (3.14) and (3.18) if we use the explicit form of the corresponding operator norm (1.5) for the case of the total variation norm. It remains to prove that condition (B) is valid for the chain X with $n = 1$ and $\rho = 1 - a(2c-1)$.

The kernel $T(x, B) = P(x, B) - aL(C \cap (B-x))$ is nonnegative by the property of the distribution F. Here and in the sequel $B - x = \{b - x \mod 1, b \in B\}$ and $\|T\| = \sup T(x, E) = 1 - ac < 1$.

Let $\mu \in \mathfrak{M}^{01} = \{\mu \in \mathfrak{M}: \mu(E) = 0, |\mu|(E) \leq 1\}$. Then $\mu P(B) = \mu T(B) + a \int \mu(dx) L(C \cap (B-x)) = \mu T(B) + a \int_B \mu(C-u)\, du$. Therefore

$$\|\mu P\| \leq \|\mu T\| + a \int |\mu(C-u)|\, du \leq 1 - ac + a\varphi(\mu), \tag{3.24}$$

where the functional $\varphi(\mu) = \int |\mu(C-u)|\, du$ is convex on $\mathfrak{M}^{01}$. Then the supremum of φ on $\mathfrak{M}^{01}$ coincides with the supremum of φ on the set $\partial\mathfrak{M}^{01} = \{(\delta_x - \delta_y)/2\}$ of extreme points of the closed convex set $\mathfrak{M}^{01}$. It is easily seen that, for $\mu = (\delta_x - \delta_y)/2$ and B, $\mu(B) \in \{-1, 0, 1\}$. Then

$$\begin{aligned}2\varphi(\mu) &= L(\{u: u + x \in C, u + y \notin C\}) + L(\{u: u + x \notin C, u + y \in C\}) \\ &= 2L((C-x) \cap (C-y)) - L(C-x) - L(C-y) \leq 2 - c - c = 2(1-c).\end{aligned}$$

Therefore, $\varphi(\mu) \leq 1 - c$ for $\mu \in \mathfrak{M}^{01}$. So from (3.24) we conclude that $\|\mu P\| \leq 1 - ac + a(1-c) = 1 - a(2c-1)$ for $\mu \in \mathfrak{M}^{01}$. Hence the condition (B) is valid.

The lower bound (3.23) of the left-hand side of (3.20) follows from the equality $(P^t - \Pi) f = (-\omega)^t f$, where $\omega = 1 - ac + a\pi^{-1} \sin \pi c$, $f(x) = \cos 2\pi x$, $\Pi f(x) = \int f\, dL = 0$, $\|f\| = 1$, so that $\|P^t f - \Pi f\| = \|P^t f\| = \omega^t$.

PROOF OF THEOREM 3.5. Let the kernel Q be such that $\varepsilon_n = \|Q^n - P^n\| < 1 - \rho$ and the kernel P satisfy condition (B) with constants n and ρ. Then this condition is fulfilled for Q with constants n and $\rho_1 = \rho + \varepsilon_n < 1$, too. Indeed, for all $\mu \in \mathfrak{M}^0 = \{\mu \in \mathfrak{M}: \mu(E) = 0\}$, we have $\|\mu Q^n\| \leq \|\mu P^n\| + \|\mu(Q^n - P^n)\| \leq (\rho + \varepsilon_n)\|\mu\|$. Since $\rho_1 < 1$, the operator Q^n is a contraction on the space $\mathfrak{M}^0$. Therefore, the operator $I - Q^n$ has a bounded inverse on $\mathfrak{M}^0$. Note that the equation $\nu = \nu Q^n$, for the invariant measure ν with $\nu(E) = 1$, is equivalent to the equation $(\nu - \pi)(I - Q^n) = -\pi(I - Q^n)$, where $\pi(I - Q^n) \in \mathfrak{M}^0$. So the boundedness of $(I - Q^n)^{-1}$ on $\mathfrak{M}^0$ implies that the measure $\nu - \pi \in \mathfrak{M}^0$ is defined uniquely. Thus the kernel Q^n has a unique invariant stochastic measure ν. Then the kernel Q also possesses this property. Since Q satisfies condition (B), a chain with the kernel Q is uniformly ergodic and aperiodic by Corollary 2.1.

We obtain from an evident identity

$$\nu - \pi = (\nu - \pi) Q^n + \pi D_n, \tag{3.25}$$

where $D_n = Q^n - P^n$ and $\nu - \pi \in \mathfrak{M}^0$, the estimate $\|\nu - \pi\| \le \rho_1\|\nu - \pi\| + \varepsilon_n\|\pi\|$, whence $\|\nu - \pi\| \le \varepsilon_n\|\pi\|/(1-\rho_1)$. The inequality (3.16) is proved.

It follows from the definition of the potential R_n, $R_n(I - P^n) = I - \Pi$, (1.8) and from the identity $D_n\Pi = (Q^n - P^n)\,\mathbb{I}\circ\pi = 0$ that $\pi D_n R_n = \pi D_n + \pi D_n R_n Q^n - \pi D_n R_n P^n$. Subtracting the last identity to (3.25), we obtain

$$\nu_1 = \nu_1 Q^n + \pi D_n R_n D_n,$$

where $\nu_1 = \nu - \pi - \pi D_n R_n$. Note that $\nu_1 \in \mathfrak{M}$ because of $R_n \in \mathcal{B}$ and, since $R_n\mathbb{I} = 0$ by the definition of R_n, $\nu_1(E) = \nu\mathbb{I} - \pi\mathbb{I} - \pi D_n R_n \mathbb{I} = 0$. Therefore, $\nu_1 \in \mathfrak{M}^0$ and $\|\nu_1\| \le \|\pi\|\varepsilon_n^2\|R_n\|/(1-\rho_1)$, where

$$\|R_n\| \le \|I - \Pi\|/(1-\rho) \le (1+\sigma)/(1-\rho)$$

by inequality (3.15) applied to the kernel P^n. Substituting the last estimate into the previous one we get (3.17).

To prove (3.18) we consider for $t \ge 1$ the identity

$$Q^{nt+n} - P^{nt+n} = \left(Q^{nt} - P^{nt}\right)Q^n + \left(P^{nt} - \Pi\right)\left(Q^n - P^n\right) + \Pi\left(Q^n - P^n\right). \quad (3.26)$$

Observe that $\mu\left(Q^s - P^s\right) \in \mathfrak{M}^0$ for all $\mu \in \mathfrak{M}$, $s \ge 1$. From (3.26) and condition (B) one can derive that the sequence $a_t = \|Q^{nt} - P^{nt}\|$ meets the inequalities

$$a_{t+1} \le \rho_1 a_t + \rho^t\varepsilon_n(1+\sigma) + \sigma\varepsilon_n, \quad (3.27)$$

where the estimate (3.14) is used for the kernel P^n.

Let the sequence $(b_t, t \ge 1)$ be recursively defined by

$$b_{t+1} = \rho_1 b_t + \rho^t\varepsilon_n(1+\sigma) + \sigma\varepsilon_n$$

and $b_1 = \varepsilon_n$. It is not hard to show by induction that $a_t \le b_t$ for $t \ge 1$. Moreover, the sequence b_t is convex for $\rho_1 = \rho + \varepsilon_n < 1$, since it is of the form $b_t = c_1 + c_2\rho^t + c_3\rho_1^t$. Therefore,

$$\sup_{t\ge 0}\|Q^{nt} - P^{nt}\| = \sup a_t \le \sup b_t \le \max\{b_1, b_\infty\} = \max\left\{\varepsilon_n, \sigma\frac{\varepsilon_n}{1-\rho_1}\right\}$$
$$= \sigma\frac{\varepsilon_n}{1-\rho-\varepsilon_n}.$$

To prove (3.18) it is necessary to estimate the norm of the right-hand side of the identity

$$Q^t - P^t = Q^s\left(Q^{nk} - P^{nk}\right) + \left(Q^s - P^s\right)P^{nk} \quad (3.28)$$

for $t = nk + s$, $0 \le s < n$. This estimate may be obtained from the inequality above for $Q^{nt} - P^{nt}$ by using in (3.28) the inclusion $\mu\left(Q^s - P^s\right) \in \mathfrak{M}^0$ for $\mu \in \mathfrak{M}$, the estimate (3.15) for the operator P^n, and the definition of the quantities q_n, $\overline{\varepsilon}_n$.

The inequalities (3.19) are obtained after the raising to the power

$$\varepsilon_t = \left\|(P + Q - P)^t - P^t\right\| \le \sum_{i=1}^{t}\binom{t}{i}\|P\|^{t-i}\|Q - P\|^i = (\varepsilon + \|P\|)^t - \|P\|^t$$
$$\le (\varepsilon + \|P\|)^{n-1} - \|P\|^{n-1}$$

for $t < n$. The theorem is proved. □

3.3. Estimates in terms of a residual kernel

In this paragraph, the following conditions are assumed:

(C) there exist $n \geq 1$, $\alpha \in m\mathcal{E}^+$, $h \in f\mathcal{E}^+$ such that $\pi h > 0$, $\alpha \mathbb{I} = 1$, $\alpha h > 0$, and the kernel $T(x, A) = P^n(x, A) - h(x)\alpha(A)$ is nonnegative;

(T) $\|T^m\| \leq \rho$ for some $m \geq 1$, $\rho < 1$.

By Theorem 2.3, condition (T) is fulfilled for all uniformly ergodic aperiodic Markov chains X that satisfy condition (C). As was shown in Remark 2.4, condition (C) is valid if the chain X is Harris recurrent or if condition (MC) holds true for the norm $\|\cdot\|$.

THEOREM 3.6. *Let conditions* (C), (T) *be fulfilled for* $n = m = 1$ *and the norm* $\|\cdot\|$ *satisfy conditions* (M1), (M2), (M3$'$). *Then for any* $t \geq 1$ *and for all* θ *from the interval*

$$1 - \frac{1-\rho}{1+\rho\sigma\omega} \equiv \theta_0 < \theta \leq 1 \tag{3.29}$$

the following inequality is valid

$$\|P^t - \Pi\| \leq \left(\rho^t + \frac{\theta^{t+2}}{\rho(\theta - \theta_0)}\right)(1+\sigma), \tag{3.30}$$

where

$$\sigma = \|\Pi\| \leq k\frac{\|\alpha\|}{1-\rho}, \qquad \omega = \min\{\omega_1, \omega_2\}, \tag{3.31}$$

$$\begin{aligned} \omega_1 &= 2\exp\left\{-\frac{(1-\pi h)\ln \alpha h}{\pi h(1-\alpha h)}\right\} - 1, \\ \omega_2 &= 2r(g)\pi T(I-T)^{-1}\mathbb{I} + 1 \leq 2\frac{\sigma\rho}{\pi h(1-\rho)\alpha h)} + 1 \end{aligned} \tag{3.32}$$

and the index $r(g)$ *is evaluated according to (3.38) below by*

$$r(g) = \inf\left\{\frac{1}{g_1}, \frac{1}{g_2} + \frac{1}{g_3}, \frac{1}{g_3} + \frac{1}{g_4}, \frac{4}{g_2} + \frac{1}{g_5}, \ldots\right\}, \tag{3.33}$$

where $g_n = \alpha T^{n-1}h$, $n \geq 1$.

REMARK 3.3. The value ω_2 is finite even in the case where the assumption $\alpha h > 0$ involved into (C) is not valid provided the distribution $g = (g_n, n \geq 1)$ is aperiodic. Thus the estimate (3.30) is meaningful for all chains X with aperiodic g.

COROLLARY 3.4. *Under conditions of Theorem* 3.6, *for all* $t > \theta_0/(1-\theta_0)$, *the following inequality holds true*

$$\|P^t - \Pi\| \leq \left(\rho^t + \theta_0^{t+2}(t+2)\frac{e}{\rho}\right)(1+\sigma). \tag{3.34}$$

PROOF OF THEOREM 3.6. First we observe that, under conditions of the theorem, the operator $T_1 = (I-T)^{-1}$ is bounded and

$$\|T_1\| = \|(I-T)^{-1}\| \leq \sum_{n\geq 0} \|T\|^n \leq \frac{1}{1-\rho}.$$

LEMMA 3.1. *The following identities are valid*

$$\alpha(I-T)^{-1} = \frac{\pi}{(\pi h)}, \qquad (I-T)^{-1}h = \mathbb{1}. \tag{3.35}$$

The proof of Lemma 3.1 follows from condition (C) and from equalities

$$\pi = \pi P = \pi T + (\pi h)\alpha, \qquad \mathbb{1} = P\mathbb{1} = T\mathbb{1} + h(\alpha\mathbb{1}).$$

Define the operator-valued functions

$$R_z = (I - zP + z\Pi)^{-1} - \Pi, \qquad T_z = (I - zT)^{-1}.$$

The functions R_z and T_z are bounded and analytical in the domain $D = \{z: |z| < 1\}$ according to Theorem 2.2 and Lemma 1.2. Moreover, the function R_z admits in this domain the representation $R_z = \sum_{n\geq 0} z^n (P^n - \Pi)$. The last series converges in $\mathcal{B}$ and $\Pi R_z = R_z\Pi = 0$.

Here and in the sequel the continuation of the norm $\|\cdot\|$ for the complex-valued measures is defined as

$$\|\mu + i\nu\| = \max\{\|\mu\|, \|\nu\|\}$$

for μ, $\nu \in \mathfrak{M}$.

LEMMA 3.2. *For all $z \in D$, the following identity holds true*

$$R_z = (I - \Pi)\left(T_z - \frac{T_z\mathbb{1} \circ \alpha T_z}{\alpha T_z\mathbb{1}}\right). \tag{3.36}$$

PROOF OF LEMMA 3.2. Let $\mu = \nu R_z$, $z \in D$. Then $\mu\Pi = 0$, $\mu(I - zP) = \nu(I - \Pi(I - zP + z\Pi)) = \nu(I - \Pi)$. Since $P = T + h\circ\alpha$, $\mu(I - zT) = \nu(I - \Pi) + z(\mu h)\alpha$. So $\mu = \nu(I - \Pi)T_z + z(\mu h)\alpha T_z$. Multiplying this equality by the function $\mathbb{1}$ and taking into account that $\mu\mathbb{1} = 0$, one can get $\mu h = -\nu(I - \Pi)T_z\mathbb{1}(z\alpha T_z\mathbb{1})^{-1}$. Substituting the last expression into the previous one we obtain (3.36). □

Note that condition (T) yields that the function on the right-hand side of (3.36) admits, along with T_z, the analytical continuation to the domain

$$\{z: |z| < \rho^{-1}\} \setminus \{z: \alpha T_z\mathbb{1} = 0\}.$$

Consider the function $a(z) = \alpha T_z\mathbb{1}$. Equalities $(1-z)\mathbb{1} = (I - zP)\mathbb{1} = (I - zT)\mathbb{1} - zh$ give us $(1-z)a(z) = 1 - g(z)$ with $g(z)$ to be the generating function of the sequence $g = \left(\alpha T^{n-1}h, n \geq 1\right)$. This sequence is a probability distribution: $g \geq 0$ because of $T \geq 0$; $\alpha \geq 0$, $h \geq 0$ by (C); and $\sum g_t = g(1) = \alpha(I - T)^{-1}h = \alpha\mathbb{1} = \mathbb{1}$ by Lemma 3.1. The first and second moment functionals of the distribution g are equal to

$$m = \sum tg_t = \sum t\alpha T^{t-1}h = \alpha(I - T)^{-2} = \frac{1}{\pi h}\pi\mathbb{1} = \frac{1}{\pi h},$$

$$m_2 = \sum t(t-1)g_t = \sum t(t-1)\alpha T^{t-1}h = 2\alpha T(I - T)^{-3}h = 2\frac{\pi T(I - T)^{-1}\mathbb{1}}{\pi h}. \tag{3.37}$$

Here we used Lemma 3.1.

To estimate the denominator $a(z) = (1 - g(z))/(1 - z)$ in (3.36) we need an auxiliary result.

Let $g = (g_n, n \geq 1)$ be an arbitrary aperiodic probability distribution. Then for some n the greatest common divisor (g.c.d.) of the set $\{i: 1 \leq i \leq n; g_i > 0\}$ is equal to the unity, so this set belongs to the class

$$\mathcal{I} = \{I \subset \mathbb{N}: \text{g.c.d. } I = 1, \sup I < \infty\}.$$

The definition of $\mathcal{I}$ yields that for all $I \in \mathcal{I}$ the set of integer-valued vectors

$$I^* = \left\{d \in \mathbb{Z}^I: \sum_{n \in I} n d_n = 1\right\}$$

is nonempty. Therefore, the following aperiodicity index

$$r(g) = \inf_{I \in \mathcal{I}} \min_{d \in I^*} \left(\sum_{i \in I} \frac{d_i^2}{g_i}\right) \tag{3.38}$$

is well defined and finite for each aperiodic distribution g. Taking in (3.38) $I = \{1\}$, $\{2,3\}$, $\{3,4\}$, $\{2,5\}, \ldots$, we obtain, in particular, the expression (3.33).

LEMMA 3.3. *Let the distribution $g = (g_n, n \geq 1)$ be aperiodic and have the moments $m = \sum n g_n$, $m_2 = \sum n(n-1) g_n$, and $r(g) \leq r$. Then*

(a) *for all $z \neq 1$, $|z| = 1$,*

$$\left|\frac{1 - g(z)}{1 - z}\right| \geq \frac{m}{\omega}, \tag{3.39}$$

where

$$\omega = \min\{\omega_1, \omega_2\},$$
$$\omega_1 = 2 \exp\left\{-\frac{m-1}{1-g_1} \ln g_1\right\} - 1, \tag{3.40}$$
$$\omega_2 = 1 + r m_2;$$

(b) *if $u_n = \sum_{t \geq 0} (g^{*t})_n$ is the renewal sequence corresponding to g and $g_1 > 0$ then*

$$\sum_{n \geq 0} |u_n - u_{n+1}| \leq \frac{\omega_1}{m},$$
$$\sum_{n \geq 0} \left|u_n - \frac{1}{m}\right| \leq \omega_1 \frac{m_2}{2m^2}. \tag{3.41}$$

REMARK 3.4. Estimate (3.39) is meaningful in the two following cases:

(a) provided the first moment is finite and g_1 is positive, $\omega_1 < \infty$;
(b) provided the second moment is finite and g is aperiodic, $\omega_2 < \infty$.

It is easy to check that in the class of all distributions with $m_2 < \infty$ and $g_1 > 0$ the indices ω_1 and ω_2 are incomparable, so both inequalities $\omega_1 < \omega_2$ and $\omega_2 < \omega_1$ are allowed.

PROOF OF LEMMA 3.3. Put $\Gamma = \{z: |z| = 1\} \setminus \{1\}$. Set $G_n = \sum_{t>n} g_t$, $a(z) = \sum z^n G_n = (1 - g(z))/(1 - z)$. It is clear that $a(1) = m$. Using the inequality $|z^n - 1| \le n|z - 1|$ that is valid on Γ we get the estimate

$$|a(z) - m| \le \sum |z^n - 1| G_n \le |z - 1| \frac{m_2}{2} = sm_2,$$

where $s = \sin(\varphi/2)$, $\varphi = \arg z$. Therefore, for $z \in \Gamma$,

$$\left| \frac{1 - g(z)}{1 - z} \right| \ge m - |a(z) - m| \ge m - sm_2. \tag{3.42}$$

Choose the set $I \in \mathcal{I}$ so that $g_i > 0$ for $i \in I$ and $d \in I^*$. Note that $\sum id_i = 1$. Using, for $n \ge 1$, $x, y \in \mathbb{R}$, the elementary inequalities $|\sin(x+y)| \le |\sin x| + |\sin y|$, $|\sin nx| \le n|\sin x|$, we give, for all α,

$$\begin{aligned} \sin^2 \alpha &= \left| \sin \left(\sum_{i \in I} id_i \alpha \right) \right|^2 \le \left(\sum_{i \in I} |d_i| \cdot |\sin i\alpha| \right)^2 \\ &= \left(\sum_{i \in I} \left| \frac{d_i}{g_i^{1/2}} \right| \left| g_i^{1/2} \sin i\alpha \right| \right)^2 \le \left(\sum_{i \in I} \frac{d_i^2}{g_i} \right) \left(\sum_{i \in I} g_i \sin^2 i\alpha \right) \\ &\le \left(\sum_{i \in I} \frac{d_i^2}{g_i} \right) \sum_{n \ge 1} g_n \sin^2 n\alpha. \end{aligned} \tag{3.43}$$

Here we used the Cauchy inequality. Evaluating the lower bound of the right-hand side with respect to $d \in I^*$ and $I \in \mathcal{I}$ we obtain (3.43) with $r(g) \le r$ instead of the first sum on the right hand side. Then, setting in (3.43) $\alpha = \varphi/2$, $\varphi = \arg z$, $s = \sin \alpha$, we arrive at

$$\begin{aligned} |1 - g(z)| &\ge \Re(1 - g(z)) = \sum_{n \ge 1} g_n (1 - \cos n\varphi) \\ &= 2 \sum_{n \ge 1} g_n \sin^2 n\alpha \ge 2 \frac{1}{r} \sin^2 \alpha = \frac{|1 - z| s}{r}. \end{aligned} \tag{3.44}$$

From (3.42) and (3.44) we conclude that the left-hand side of (3.39) is not less than

$$\max \left\{ \frac{s}{r}, m - sm_2 \right\}, \tag{3.45}$$

where $s = \sin(\arg z/2)$, $s \in [0, 1]$. Note that for all $z \in \Gamma$ expression (3.45) is not less than its minimal value with respect to $s \in [0, 1]$. This minimum is attained at the point $t = m(m_2 + r^{-1})^{-1}$ and is equal to $m/(1 + rm_2)$. So, estimate (3.39) is proved for $\omega = \omega_2$.

To prove (3.39) for $\omega = \omega_1$ we introduce G_n and $a(z)$ as above and consider the sequence

$$F_n = \prod_{i=1}^{n}(1 - G_i), \qquad F_0 = 1, \qquad f_n = G_n F_{n-1},$$

for $n \geq 1$, $f_0 = 1$. We check by induction that $f_1 + \cdots + f_n = 1 - F_n$ for $n \geq 1$. Define the sequence $h_n = \sum_{i=0}^{n} f_i G_{n-i} - G_n$ for $n \geq 1$ and $h_0 = 1$. Note that the sequence h_n is nonnegative. Indeed, since f_n is nonnegative and G_n is nonincreasing,

$$\begin{aligned} h_n &= f_n G_0 + f_{n-1} G_2 + \cdots f_1 G_{n-1} - G_n \geq f_n + (f_1 + \cdots + f_{n-1}) G_n - G_n \\ &= f_n + (1 - F_{n-1}) G_n - G_n = 0 \end{aligned}$$

for $n \geq 2$ and $h_0 = 0$, $h_1 = f_1 - G_1 = 0$.

Denote by $h(z)$ and $f(z)$ the generating functions of the sequences h_n and f_n respectively. Then $|h(z)| \leq \sum h_n = h(1)$ for $z \in \Gamma$. Since

$$h(z) = f(z)a(z) - a(z) + 1, \tag{3.46}$$

$h(1) = 1 - (1 - f(1))a(1)$. Here

$$f(1) = \sum f_n = \lim(1 - F_n) = 1 - F \leq 1, \qquad F = \prod_{n \geq 1}(1 - G_n), \qquad a(1) = m.$$

Then $h(1) = 1 - mF \leq 1$. Moreover, for $z \in \Gamma$,

$$\begin{aligned} |(1 - f(z))a(z)| &= |1 - h(z)| \geq 1 - |h(z)| \geq 1 - h(1) = mF, \\ |1 - f(z)| &\leq 1 + |f(z)| \leq 1 + f(1) = 2 - F. \end{aligned}$$

So, for all $z \in \Gamma$,

$$\left| \frac{1 - g(z)}{1 - z} \right| = |a(z)| \geq \frac{mF}{2 - F}. \tag{3.47}$$

Estimate (3.39) for $\omega = \omega_1$ immediately follows from (3.47) if we take into account the following inequality

$$F \geq \exp\left\{ \frac{m - 1}{1 - g_1} \ln g_1 \right\} = \frac{2}{1 + \omega_1}. \tag{3.48}$$

Here the functional $F = F(G)$ of the sequence G is defined by

$$F(G) = \prod_{n \geq 1}(1 - G_n).$$

We set

$$G \in \Phi = \left\{ (G_n, n \geq 1) : 0 \leq G_n \leq 1 - g_1, \sum G_n \leq m - 1 \right\}.$$

It is easy to check that the set Φ is convex and the function $\ln F(G)$ is concave on Φ. Thus its infimum is attained at the boundary of Φ. The latter contains all sequences G such that $G_n \in \{0, 1-g_1\}$ for $n \geq 1$ with the exception at most one number n. Evaluating $F(G)$ for such sequences G we get (3.48). The part (a) of the lemma is proved.

To prove the part (b), set $u(z) = \sum z^n u_n$, $\bar{a}(z) = \sum z^n \overline{G}_n$, where $\overline{G}_n = \sum_{t \geq n} G_t$. We obtain from (3.46) that

$$(1-z)u(z) = \frac{1-z}{1-g(z)} = \frac{1}{a(z)} = \frac{1-f(z)}{1-h(z)},$$
$$u(z) - \frac{1}{m(1-z)} = \frac{1-g(z)-m(1-z)}{m(1-z)(1-g(z)} = \frac{\bar{a}(z)}{ma(z)} = \frac{\bar{a}(z)(1-f(z)}{m(1-h(z)},$$

where the sequences f_n and h_n are as above. This representation implies

$$\begin{aligned} u_n - u_{n-1} &= H_n - (f * H)_n, \\ u_n - \frac{1}{m} &= \frac{1}{m} H * (G - G * f)_n \end{aligned} \tag{3.49}$$

where the sequence $H_n = \sum_{t \geq 0} (h^{*t})_n$ is nonnegative and $\sum H_n = (1 - \sum h_n)^{-1} = 1/mF < \infty$. Since $\sum f_n = 1 - F$ and $\sum \overline{G}_n = m_2/2$, (3.49) yields

$$\sum |u_n - u_{n-1}| \leq \sum H_n \left(1 + \sum f_n\right) = \frac{2-F}{mF},$$
$$\sum \left| u_n - \frac{1}{m} \right| \leq \frac{1}{m} \sum \overline{G}_n \left(1 + \sum f_n\right) \sum H_n \leq \frac{m_2}{m} \frac{2-F}{2mF}.$$

Putting (3.48) into these inequalities we get (3.41). The lemma is proved. □

Now we come back to the proof of the theorem. From (3.32) we obtain ω_1 and ω_2 by substitution of expressions for moments (3.37) into (3.40). Therefore, Lemma 3.3 implies

$$\inf_{z \in \Gamma} |a(z)| \geq \frac{1}{\omega(\pi h)}, \tag{3.50}$$

where $a(z) = \alpha T_z \mathbb{I}$ and the index ω is defined by (3.32). Now we take $r \in (1, \rho^{-1})$ and write down, for $|z| = 1$,

$$\begin{aligned} |a(rz) - a(z)| &= (r-1)|z\alpha T_{rz} T T_z \mathbb{I}| = (r-1) \left| \sum_{i,j \geq 0} z^{i+j} r^i T^{i+j+1} \mathbb{I} \right| \\ &\leq (r-1) \sum r^i T^{i+j+1} \mathbb{I} = (r-1) \alpha T_1 T_r \mathbb{I} = \frac{r-1}{\pi h} \pi T_r T \mathbb{I} \\ &= (r-1) \frac{\sigma \rho}{\pi h (1 - \rho r)}. \end{aligned} \tag{3.51}$$

Here an evident estimate $\|T_r\| \leq (1-\rho r)^{-1}$ and Lemma 3.1 were used.

Since $|a(rz)| \geq |a(z)| - |a(rz) - a(z)|$, we obtain by using (3.50) and (3.51)

$$|a(rz)| \geq \frac{1}{(\pi h)\omega} \left(1 - (r-1)\frac{\rho \sigma \omega}{1 - \rho r}\right) = (1 - \theta_0 r) \frac{1 - \rho}{\omega(\pi h)(1 - \rho r)(1 - \theta_0} > 0 \tag{3.52}$$

for all $1 \le r < \theta_0^{-1}$ and z with $|z| = 1$.

Condition (M3′) for the norm $\|\cdot\|$ gives us that $\|\mu\| = \max\{\||\Re\mu|\|, \||\Im\mu|\|\}$ for any complex-valued measure μ. Since $|\Re\alpha T_{rz}| = |\sum \Re r^n z^n \alpha T^n| \le \sum r^n \alpha T^n = \alpha T_r$ and $|\Im \alpha T_{rz}| \le \alpha T_r$,

$$\|\alpha T_{rz}\| \le \|\alpha T_r\| = \left\|\alpha T_1 (I - (r-1) T T_1)^{-1}\right\| \\ \le \|\alpha T_1\| \left(1 - (r-1)\rho(1-\rho)^{-1}\right)^{-1} = \|\pi\| \frac{1-\rho}{(1-\rho r)\pi h} \tag{3.53}$$

for $|z| = 1$ in view of (M3').

Since T_z is analytical for $|z| < \rho^{-1}$, the function R_z admits the analytical continuation to the domain $\{z : |z| < \theta_0^{-1}\}$ because of (3.36) and (3.52). Applying the Cauchy formula to R_z and T_z, [see Dunford and Schvartz (1953), Chapter 7.3], we obtain

$$P^t - \Pi = \frac{1}{2\pi i} \int_{|z|=1} \frac{1}{z^{t+1}} ((I - \Pi) T_z + R_z + (\Pi - I) T_z)\, dz \\ = (I - \Pi) T^t + \frac{1}{2\pi i} \int_{|z|=r} \frac{1}{z^{t+1}} (R_z + (\Pi - I) T_z)\, dz \\ = (I - \Pi) \left[T^t - \frac{r^{-t}}{2\pi i} \int_{|z|=1} \frac{1}{z^{t+1}} \frac{T_{rz} \mathbb{1} \circ \alpha T_{rz}}{a(rz)}\, dz \right] \tag{3.54}$$

for all $r < \theta_0^{-1}$. The integral in (3.54) is estimated by (3.52), (3.53), and the inequality $\|T_{rz}\| \le (1 - \rho r)^{-1}$. After simple calculations we arrive at the inequality (3.30) substituting $r = \theta^{-1}$ for $\theta > \theta_0$.

The estimate (3.34) is derived from (3.30) by substituting $\theta = \theta_0 (t+2)/(t+1)$ with $\theta \le 1$ for $t > \theta_0 (1 - \theta_0)^{-1}$. The theorem is completely proved. □

The following result is valid for all norms (not necessarily compatible with the order structure in $\mathfrak{M}^+$).

THEOREM 3.7. *Let conditions* (C) *and* (T) *be valid for some* n, α, h, m, ρ, *and let the norm* $\|\cdot\|$ *satisfy condition* (M1) *with* $\sigma = \|\Pi\| < \infty$. *Then for all* $t \ge 1$ *and* θ *from the interval*

$$1 - \frac{1-\rho}{1 + \sigma b_m \omega} = \theta_m < \theta \le 1 \tag{3.55}$$

the following inequality

$$\left\|P^t - \Pi\right\| \le \left(\rho^s + \theta^s \frac{c_m}{(\theta - \theta_m)(\theta - \rho)}\right) b_m p_m \tag{3.56}$$

holds true, where

$$s = \left[\frac{[t/n]}{m}\right], \qquad b_m = \max_{i<n} \left\|T^i\right\|, \\ p_n = \max_{j<n} \left\|P^j - \Pi\right\|, \qquad c_m = m^2 (\pi h) \frac{\|\alpha\|}{\|\pi\|}, \tag{3.57}$$

and the constant ω is defined by (3.32).

PROOF. Since $(I-T)^{-1} = \left(I+\cdots+T^{m-1}\right)(I-T^m)^{-1}$ for $\|T^m\| < 1$, the operator $I-T$ has a bounded inverse. Moreover, the functions R_z and T_z that appeared in the proof of Theorem 3.6 are analytical in a neighborhood of zero, since $P \in \mathcal{B}$, $\Pi \in \mathcal{B}$, $T \in \mathcal{B}$. Thus the assertions of Lemmas 3.1 and 3.2 are valid. The estimates given below show that the right-hand side of (3.36) admits the analytical continuation to the domain $|z| < r$. As above we obtain representation (3.54) with the kernel P^n instead of P. Note that only the first degree of P^n is involved in condition (C).

Put $l = [t/n]$, so $t = nl + j$, $0 \le j < n$. Multiplying (3.54) by P^j and using the estimate $\left\|P^j - \Pi\right\| \le p_n$ we obtain the inequality

$$\left\|P^t - \Pi\right\| \le p_n\left(\left\|T^l\right\| + \frac{1}{r^l}\sup_{|z|=1}\left|\frac{1}{a(rz)}\right| \|T_{rz}\mathbb{1}\|\cdot\|\alpha T_{rz}\|\right). \tag{3.58}$$

Set $s = [l/m]$ and substitute $r = \theta^{-1/m}$ into (3.58). To derive (3.56) we need to estimate the norms from (3.58). Using the following relation

$$\begin{aligned}(r-1)\pi T_r T\mathbb{1} &\le (r-1)\sigma\|T_r T\| \le (r-1)\sigma\left\|\left(T+\cdots+r^{m-1}T^m\right)\left(I-r^mT^m\right)^{-1}\right\| \\ &\le \frac{r-1}{1-\rho r^m}\sigma\left(b_m\left(1+\cdots+r^{m-2}\right)+\rho r^{m-1}\right) \\ &= \frac{\sigma}{1-\rho r^m}\left(r^{m-1}(b_m-\rho) - b_m + \rho r^m\right) \\ &\le \frac{\sigma}{1-\rho r^m}\left(r^m(b_m-\rho) - b_m + \rho r^m\right) = \frac{\sigma}{\theta-\rho}b_m(1-\theta)\end{aligned}$$

instead of the last inequality in (3.51) and estimating the difference $|a(rz) - a(z)|$ according to (3.51) we conclude from (3.50) that for $|z| = 1$

$$\begin{aligned}|a(rz)| &\ge \frac{1}{\omega(\pi h)} - \sigma b_m\frac{1-\theta}{\theta-\rho}\pi h \\ &= (\theta-\theta_m)\frac{1-\rho}{\omega\pi h(1-\theta_m)(\theta-\rho)}.\end{aligned} \tag{3.59}$$

To estimate the norms of αT_{rz} and $T_{rz}\mathbb{1}$ we use the inequality

$$\|T_{rz}\| = \left\|\left(I+\cdots+(rzT)^{m-1}\right)\left(I-(rzT)^m\right)^{-1}\right\| \le mb_m\frac{r^m}{1-r^m\rho} = \frac{mb_m}{\theta-\rho}. \tag{3.60}$$

Finally, for $l = ms+i$, $0 \le i < m$, an evident estimate $\left\|T^l\right\| \le \|T^{ms}\|\cdot\left\|T^i\right\| \le \rho^s b_m$ is valid. Substituting (3.59) into (3.58), using twice (3.60) in (3.58), and the latter estimation we get (3.56). The theorem is proved. $\square$

THEOREM 3.8. *Let conditions* (C) *and* (T) *be satisfied for $n = m = 1$ and the norm $\|\cdot\|$ satisfy condition* (M1) *with $\sigma = \|\Pi\| < \infty$. Then any stochastic transition kernel $Q = P + D$ from a neighborhood*

$$\|Q - P\| = \varepsilon < \varepsilon_0 = \frac{1-\rho}{1+\sigma\rho} \tag{3.61}$$

has a unique invariant stochastic measure ν and

$$\|\nu - \pi\| \leq \varepsilon \frac{\|\pi\|}{\varepsilon_0 - \varepsilon}, \tag{3.62}$$

$$\|\nu - \pi - \pi D R_T (I - \Pi)\| \leq \varepsilon^2 \frac{\|\pi\|}{\varepsilon_0 - \varepsilon}, \tag{3.63}$$

where $R_T = (I - T)^{-1}$ and the residual kernel T is defined in condition (C).

PROOF. Let $\nu = \nu Q$, $\nu \mathbb{I} = \mathbb{I}$. Then condition (C) yields $\nu = \nu D + \nu T + (\nu h)\alpha$, whence $\nu = \nu D R_T + (\nu h)\alpha R_T = \nu D R_T + (\nu h)\pi$ because of Lemma 3.1. To evaluate νh we note that $\mathbb{I} = \nu \mathbb{I} = \nu D R_T \mathbb{I} + \nu h$. Therefore $\nu = \nu D R_T + (1 - \nu D R_T)\pi = \nu D R_T (I - \Pi) + \pi$. Since $\pi \in \mathfrak{M}$, $D\Pi = 0$, and

$$\|D R_T (I - \Pi)\| = \|D(I - \Pi) + D T R_T (I - \Pi)\| \leq \varepsilon + \varepsilon \frac{\rho}{1-\rho}(1+\sigma) = \frac{\varepsilon}{\varepsilon_0} < 1, \tag{3.64}$$

$\nu \in \mathfrak{M}$ and the following representation

$$\nu = \pi (I - D R_T (I - \Pi))^{-1}. \tag{3.65}$$

is valid. Making all conclusions in the reverse order, we make sure that the measure ν in (3.65) is invariant for Q and $\nu \mathbb{I} = 1$.

Estimates (3.62) and (3.63) immediately follow from (3.65) and (3.64) by using the Neuman decomposition

$$\nu = \pi + \pi A (I - A)^{-1} = \pi + \pi A + \pi A^2 (I - A)^{-1}$$

with the operator $A = D R_T (I - \Pi)$. The theorem is proved. □

THEOREM 3.9. *Let conditions* (C) *and* (T) *be satisfied for $n = m = 1$ and the norm $\|\cdot\|$ satisfy condition* (M1) *with $\sigma = \|\Pi\| < \infty$. Then for any stochastic transition kernel $Q = P + D$ from a neighborhood*

$$\|Q - P\| = \varepsilon < \varepsilon_1 = \frac{1-\rho}{1+\rho\sigma\omega} \tag{3.66}$$

the following inequality

$$\sup_{t \geq 0} \left\|Q^t - P^t\right\| \leq \varepsilon \frac{2\sigma + 1 + \sigma(\varepsilon_1 - \varepsilon)}{(\varepsilon_0 - \varepsilon)(\varepsilon_1 - \varepsilon)} \tag{3.67}$$

holds true where the constants ε_0 and ω are defined by (3.61) *and* (3.32) *respectively.*

PROOF. Let us introduce the operator-valued functions $P_z = (I - zP)^{-1}$ and $Q_z = (I - zQ)^{-1}$ that are analytical in a neighborhood of zero. Denote by $N = \mathbb{I} \circ \nu$ the stationary projector of the kernel Q, where, according to Theorem 3.8, ν is the unique invariant stochastic measure of Q.

LEMMA 3.4. *For small* $|z|$,

$$Q_z - P_z = \frac{1}{1-z}(N - \Pi) + z(I - \Pi)P_z D Q_z + (N - \Pi)Q_z. \tag{3.68}$$

PROOF. The proof of Lemma 3.4 consists in checking a simple identity

$$(I-zP)-(I-zQ) = \frac{1}{1-z}(I-zP)(N-\Pi)(I-zQ)+z(I-\Pi)D+(I-zP)(N-\Pi).$$

Then we multiply it on the left by the operator P_z that is bounded for small $|z|$ and on the right by Q_z. □

Define the following functions

$$T_z = (I - zT)^{-1}, \qquad S_z = (I - zT - zD)^{-1}, \qquad a(z) = \alpha T_z \mathbb{I},$$
$$A_z = T_z - \frac{1}{a(z)} T_z \mathbb{I} \circ \alpha T_z, \qquad b(z) = \alpha S_z \mathbb{I}, \qquad B_z = S_z - \frac{1}{b(z)} S_z \mathbb{I} \circ \alpha S_z.$$

The functions T_z, S_z, $a(z)$, and $b(z)$ are analytical in a neighborhood of the unit circle, since $\|T\| \le \rho < 1$ and $\|T + D\| \le \rho + \varepsilon < 1$ for all ε satisfying (3.36).

Note that $GP_z = GA_z$ for any operator G with $G\mathbb{I} = 0$. Indeed, by Lemma 3.2

$$GR_z = G(I - \Pi)A_z = GA_z - (G\mathbb{I}) \circ \pi A_z = GA_z.$$

Then $GR_z = G\left(P_z - (1-z)^{-1}\Pi\right) = GP_z$. Analogously, $GQ_z = GB_z$, since we did not use the fact that the kernel T is non-negative while proving (3.36) in Lemma 3.2. Also $Q = P + D = (T + D) + h \circ \alpha$. Now from (3.68) we get

$$\begin{aligned} \sum z^t \left(Q^t - P^t - N + \Pi\right) &= Q_z - P_z - \frac{1}{1-z}(N - \Pi) \\ &= z(I - \Pi)A_z D B_z + (N - \Pi)B_z, \end{aligned} \tag{3.69}$$

since the actions of the operators $I - \Pi$, D, and $N - \Pi$ on the function $\mathbb{I}$ equal to zero. It is showed below that the functions $a(z)$ and $b(z)$ are separated from zero in a neighborhood of the unit circle. Thus A_z and B_z are analytical in this neighborhood and by the Cauchy formula

$$Q^t - P^t - N + \Pi = \frac{1}{2\pi i}\int_{|z|=1} \frac{1}{z^{t+1}}\left[(I - \Pi)A_z D + N - \Pi\right] B_z\, dz.$$

Therefore

$$\sup_t \left\|Q^t - P^t\right\| \le \|N - \Pi\| + \sup_{|z|=1}\left[\left((1+\sigma)\varepsilon\|A_z\|_0 + \|N - \Pi\|\right)\|B_z\|_0\right], \tag{3.70}$$

where the reduced operator norm $\|C\|_0 = \sup\left(\|\mu C\|, \mu\mathbb{I} = 0, \|\mu\| \le 1\right)$ is used, since the operators A_z and B_z in the representation above are multiplied from the left by the operators $I - \Pi$, D, $N - \Pi$ that take values in the space $\mathfrak{M}^0 = \{\mu \in \mathfrak{M}\colon \mu\mathbb{I} = 0\}$.

To estimate $\|A\|_0$ we note that $|a(z)| \geq 1/\omega(\pi h)$ for $|z| = 1$ due to (3.50). Moreover,

$$\|T_z\|_0 \leq \|T_z\| \leq \frac{1}{1-\rho},$$

$$\begin{aligned} \|T_z\mathbb{I} \circ \alpha T_z\|_0 &= \|T_z\mathbb{I}\|_0 \cdot \|\alpha T_z\| = \|\mathbb{I} + zTT_z\mathbb{I}\|_0\|\alpha T_z\| \\ &= \|zTT_z\mathbb{I}\|_0\|\alpha T_z\| \leq \|\mathbb{I}\|\frac{\rho}{1-\rho}\|\alpha T_1\| = \sigma\frac{\rho}{(1-\rho)\pi h}. \end{aligned} \tag{3.71}$$

Therefore, for $|z| = 1$,

$$\|A_z\|_0 \leq \frac{1}{1-\rho} + \sigma\omega\frac{\rho}{1-\rho} = \frac{1}{\varepsilon_1}. \tag{3.72}$$

Similar estimates are valid for S_z, too,

$$\|S_z\|_0 \leq \frac{1}{1-\rho-\varepsilon},$$

$$\|S_z\mathbb{I}\|_0 \leq \left\|(I - zT_zD)^{-1}T_z\mathbb{I}\right\| \leq (1 - \varepsilon\|T_z\|)^{-1}\,\|TT_z\mathbb{I}\|_0 \leq \|\mathbb{I}\|\frac{\rho}{1-\rho-\varepsilon},$$

$$\|\alpha S_z\| = \left\|\alpha T_z(I - zDT_z)^{-1}\right\| \leq \frac{\|\alpha T_z\|}{1-\varepsilon\|T_z\|} \leq \|\pi\|\frac{1-\rho}{(1-\rho-\varepsilon)\pi h}. \tag{3.73}$$

Furthermore,

$$\begin{aligned} |b(z) - a(z)| &= |\alpha(S_z - T_z)\mathbb{I}| = |\alpha T_zDS_z\mathbb{I}| \leq \|\alpha T_z\|\varepsilon\|S_z\mathbb{I}\| \\ &\leq \|\pi\|\varepsilon\|\mathbb{I}\|\frac{\rho}{(1-\rho-\varepsilon)\pi h}. \end{aligned}$$

Substituting this inequality into the estimate $|b| \geq |a| - |b-a|$ and using (3.60) we obtain for $|z| = 1$ the lower estimate

$$|b(z)| \geq \frac{1}{\omega(\pi h)} - \sigma\varepsilon\frac{\rho}{(1-\rho-\varepsilon)\pi h} = (1-\rho)\frac{\varepsilon_1 - \varepsilon}{\omega(\pi h)\varepsilon_1(1-\rho-\varepsilon)} > 0.$$

Applying the last inequality to $\|B_z\|_0$ we derive for $\varepsilon < \varepsilon_1$ and $|z| = 1$ that

$$\begin{aligned} \|B_z\|_0 &\leq \frac{1}{1-\rho-\varepsilon} + \omega(\pi h)\frac{\varepsilon_1}{\varepsilon - \varepsilon_1}\,\frac{1-\rho-\varepsilon}{1-\rho}\,\sigma\frac{\rho}{(\pi h)^2}\,\frac{1-\rho}{(1-\rho-\varepsilon)^2} \\ &= \frac{\varepsilon_1(1+\rho\sigma\omega) - \varepsilon}{(1-\rho-\varepsilon)(\varepsilon_1 - \varepsilon)} = \frac{1}{\varepsilon_1 - \varepsilon}. \end{aligned} \tag{3.74}$$

It remains to substitute (3.72) and (3.74) into (3.70) and to estimate $\|N - \Pi\| = \|\mathbb{I} \circ (\nu - \pi)\|$ by inequality (3.62). It is clear that $\varepsilon < \varepsilon_1$ implies $\varepsilon < \varepsilon_0$, since $\omega_1 > 1$. The theorem is proved. □

REMARK 3.5. Similar estimates can be derived from (3.62), (3.63), and (3.67) in the case where condition (T) is fulfilled for arbitrary n and m. To do this one can apply, as in Theorem 3.6, inequalities of Theorems 3.8 and 3.9 to the kernel P^n and then use in expressions (3.71), (3.72), (3.73) the following estimate

$$\|R_T\| \leq \left\|\left(I + \cdots + T^{m-1}\right)\left(I - T^m\right)^{-1}\right\| \leq \frac{mb_m}{1-\rho}$$

instead of the inequality $\|R_T\| \leq (1-\rho)^{-1}$. The estimate for the deviation $Q^t - P^t$ in terms of $Q^{nk} - P^{nk}$ may be derived from (3.28) in the same way as that in Theorem 3.5.

CHAPTER IV

ASYMPTOTIC EXPANSIONS IN ERGODIC AND STABILITY THEOREMS

We obtain in this chapter remainder terms in ergodic and stability theorems with an order of rest terms. Observe that corresponding bounds are uniform on a wide class of uniformly ergodic Markov chains.

In the sequel, we assume that a chain $X = (X_t, t \geq 0)$ has a unique invariant stochastic measure π satisfying condition (P) and that condition (M1) is fulfilled for the norm $\|\cdot\|$.

4.1. Expansions in ergodic theorems

Define the following ergodic index, that depends only on the transition kernel P and on the norm $\|\cdot\|$ on $\mathfrak{M}$

$$\omega(P) = \inf_{n\geq 1} \inf_{K\in\mathcal{K}} (\|P^n - K\|)^{1/n}, \tag{4.1}$$

where $\mathcal{K}$ is the class of compact operators in $\mathcal{B}$.

REMARK 4.1. Theorem 2.2 with (M1)–(M3) and (C) imply that a chain X is uniformly ergodic if and only if $\omega(P) < 1$. Furthermore, the estimate $\omega(P) \leq \rho^{1/mn}$ is valid, where the constants ρ, m, and n are involved into conditions (Cd) and (T). These conditions are fulfilled by Theorem 2.4.

Denote by $P_z = (z - P)^{-1}$ the resolvent of the operator $P \in \mathcal{B}$ and consider for $0 \leq \rho < 1$ the set

$$D(\rho) = \{\xi : \rho \leq |\xi| < 1, P_\xi \notin \mathcal{B}\}. \tag{4.2}$$

THEOREM 4.1. *Let $\omega(P) < 1$. Then*

(a) *for all $\rho > \omega(P)$ the set $D(\rho)$ is finite;*

(b) *the inclusion $\xi \in D(\rho)$ holds true if and only if there exist a natural $n(\xi)$ and a family of bounded operators $(C_k(\xi), 0 \leq k \leq n(\xi))$ such that $C_{n(\xi)}(\xi) \neq 0$ and the operator-valued function*

$$\Delta_\xi(z) = P_z - \sum_{k=0}^{n(\xi)} (z - \zeta)^{-k} C_k(\xi) \tag{4.3}$$

is analytical in a neighborhood of ξ and $\Delta_\xi(\xi) = 0$;

(c) *the index $n(\xi)$ in (4.3) equals the minimal n such that $\mathfrak{M}(P - \xi)^{n+1} = \mathfrak{M}(P - \xi)^n$ or, equivalently, $\ker(P - \xi)^{n+1} = \ker(P - \xi)^n$;*

(d) *for $n(\xi) = n$, the operator $C_n(\xi)$ is a proper projector for the operator $P - \xi$, that is,*

$$C_n(\xi)(P - \xi) = (P - \xi)C_n(\xi) = 0, \qquad C_n^2(\xi) = C_n(\xi), \tag{4.4}$$

and the operator $C_0(\xi)$ is equal to the zero extension of the operator

$$\begin{gathered} C(\xi) : \mathfrak{M}(P - \xi)^n \to \mathfrak{M} \bmod \ker(P - \xi)^n, \\ C(\xi) = \left[C_n(\xi) - (P - \xi)^{n-1}\right] \left[(P - \xi)^n\right]^{-1}, \end{gathered} \tag{4.5}$$

and for all $1 \le k \le n$,

$$C_k(\xi) = C_0(\xi)(P-\xi)^k + (P-\xi)^{k-1}. \tag{4.6}$$

REMARK 4.2. It follows from Theorem 4.1 that the point ξ belongs to the domain

$$D(\omega) = \bigcup_{\rho > \omega(P)} D(\rho) \tag{4.7}$$

if and only if the operator $P - \xi$ has an annihilator in $\mathfrak{M}$ or, equivalently, in the dual space $\mathfrak{N}$. By (4.4) these spaces of annihilators are $\mathfrak{M}C_n(\xi)$ and $C_n(\xi)\mathfrak{N}$, respectively.

Observe also that, for $\rho = \omega(P)$, assertion (a) of the theorem can be broken, so the set $D(\omega)$ may be infinite. Precisely this situation is presented in the following example.

EXAMPLE 4.1. Let $E = \mathbb{Z}_+$, the chain X be given by the matrix

$$P = (P_{ij}, i, j \in E),$$

where $P_{00} = 1$, $p_{i0} = q_i$, $p_{ii} = p_i = 1 - q_i < 1$, and let the norm of a measure $\mu = (\mu_j)$ be equal to $\|\mu\| = \sum |\mu|$. Using (4.1) we evaluate

$$\omega(P) = \inf_{n \ge 0} \sup_{t > n} p_t = \limsup_{t \to \infty} p_t.$$

Therefore the chain X is uniformly ergodic provided that $\limsup p_n < 1$. The spectrum of P consists of the points $\{p_n, n \ge 1\} \cup \{1\}$. Thus $D(\rho) = \{p_n : p_n \ge \rho\}$ and the set $D(\omega)$ is infinite if, for example, p_n strictly decrease to the limit $\omega(P)$.

PROOF OF THEOREM 4.1. Let $\rho > \omega(P)$. Then by (4.1) there exist $n \ge 1$ and $K \in \mathcal{K}$ such that $\|P^n - K\| \le \rho^n$. Therefore, $\mathcal{B}$-valued function $Q_z = (z^n - P^n + K)^{-1}$ is analytical for $|z| > \rho$. Since $Q_z\left(z^{n-1} + \cdots + P^{n-1}\right) = Q_z(z^n - P^n)P_z = Q_z\left(Q_z^{-1} - K\right)P_z = (I - Q_z K)P_z$, the singularities of P_z in the domain $|z| > \rho$ are contained in the set of singular points of the function $T_z = (I - Q_z K)^{-1}$. By Theorem 7.1.9 from Dunford and Schvartz (1953), the latter set is finite, so the part (a) is proved.

Since all singular points of T_z are isolated and have finite multiplicity because the operator $Q_z K$ is compact, the function T_z is finite and meromorphic for $|z| > \rho$. Then the same property holds true also for P_z, so assertion (b) is also proved.

Note that, by the definition of the resolvent $P_z(z - P) = (z - P)P_z = I$, the function $\Delta_\xi(z)$ in (4.3) is analytic in a neighborhood of ξ and equals zero at the point $z = \xi$ if and only if the following statements are valid

$$\begin{aligned} C_n(\xi)P(\xi) &= P(\xi)C_n(\xi) = 0, \qquad C_n(\xi) \ne 0, \\ C_{k-1}(\xi)P(\xi) &= P(\xi)C_{k-1}(\xi) = C_k(\xi), \qquad 1 < k \le n, \\ C_0(\xi)P(\xi) &= P(\xi)C_0(\xi) = I + C_1(\xi), \end{aligned} \tag{4.8}$$

where $P(\xi) = P - \xi$ and $n = n(\xi)$.

Parts (c) and (d) immediately follow from (4.8). Let $\mu \in \mathfrak{M}P^n(\xi)$, $\mu = \nu P^n(\xi)$. Then $\mu = \nu\left(C_0(\xi)P^2(\xi) - C_1(\xi)P(\xi)\right)P^{n-1}(\xi) = \nu C_0(\xi)P^{n+1}(\xi) - \nu C_0(\xi)P^n(\xi) \in \mathfrak{M}P^{n+1}(\xi)$, since $C_1(\xi)P^n(\xi) = C_2(\xi)P^{n-1}(\xi) = \cdots = C_n(\xi)P(\xi) = 0$ and $I = C_0(\xi)P(\xi) - C_1(\xi)$. In the same way we prove the second assertion in (c). Furthermore, it follows from the Fredholm property of $P(\xi)$, [see Kato (1966), Theorem 4.5.26], that the subspaces $\mathfrak{M}P^n(\xi)$ and $\ker P^n(\xi)$ are closed and the codimension of the first one is equal to the dimension of the second.

The first identity in (4.4) coincides with the first equality in (4.8). The second identity in (4.4) follows from the resolvent identity $P_z - P_u = -(z-u)^{-1}P_zP_u$ in a neighborhood of ξ.

Relation (4.6) is obtained consequently by evaluation of $C_k(\xi)$ from (4.8).

Now let the operators $C_n(\xi)$ and $C_0(\xi)$ be obtained from (4.4) and (4.5) respectively. Note that $C_0(\xi)$ is well defined, since the spaces $\mathfrak{M}P^n(\xi)$ and $\ker P^n(\xi)$ are closed. By the closed graph theorem, [see Kato (1966), Theorem 3.5.20], the operator $C_0(\xi)$ is bounded. Furthermore, $C_0(\xi)$ commutes with $P(\xi)$, since $C_n(\xi)$ possesses this property and the subspaces $\mathfrak{M}P^n(\xi)$ and $\ker P^n(\xi)$ are invariant for $P(\xi)$. Then relations (4.8) immediately follow from (4.4), (4.5), and (4.6). Since the solution to (4.8) is unique because the main part of the finite-meromorphic function P_z is unique in a neighborhood of a singular point ξ, the operators $C_k(\xi)$ defined in (d) coincide with coefficients from (4.3). The theorem is proved. □

THEOREM 4.2. *Let a chain X be uniformly ergodic in a norm $\|\cdot\|$ and aperiodic, so $\omega(P) < 1$. Then for all $\rho \in (\omega(P), 1)$ the following asymptotic expansion holds true*

$$P^t - \Pi = \sum_{\xi \in D(\rho)} \sum_{k=0}^{n(\xi)-1} t(t-1)\cdots(t-k+1)\xi^{t-k}\frac{C_{k+1}(\xi)}{k!} + O_\rho(t) \tag{4.9}$$

as $t \to \infty$ where $\Pi = \mathbb{I} \circ \pi$ is the stationary projector for the kernel P, the finite set $D(\rho)$ is defined by (4.2) and described in Theorem 4.1, the index $n(\xi)$ and the operators $C_k(\xi)$ are given by (4.4), (4.5), (4.6), and an operator-valued function $O_\rho(t)$ is such that $\|O_\rho(t)\| = O\left((\rho - \varepsilon)^t\right)$ for some $\varepsilon > 0$ as $t \to \infty$.

Note that the operators $O_\rho(t)$ in (4.9) converge to zero in the norm very quickly compared with all other summands on the right-hand side of (4.9), since $|\xi| \geq \rho$ for all $\xi \in D(\rho)$. It was showed above that, for $\rho = \omega(P)$, relation (4.9) may be not satisfied. So all terms of the exponential asymptotics of $P^t - \Pi$, that have orders $\omega(P)$ and higher, are not exponential polynomials. This property can be easily obtained from Example 4.1.

PROOF. The proof of Theorem 4.2 immediately follows from Theorem 4.1 and the integral Cauchy formula

$$P^t = \frac{1}{2\pi i}\int_{|z|=1+\varepsilon} z^t P_z\,dz = \frac{1}{2\pi i}\int_{|z|=1+\varepsilon} z^t M_z\,dz + \frac{1}{2\pi i}\int_{|z|=\rho-\varepsilon} z^t \Lambda_z\,dz, \tag{4.10}$$

where

$$M_z = \frac{1}{z-1}\Pi + \sum_{\xi \in D(\rho)} \sum_{k=1}^{n(\xi)} \frac{1}{(z-\xi)^k} C_k(\xi)$$

is the main part of the meromorphic function P_z in the domain $|z| > \rho - \varepsilon$ and the function $\Lambda_z = P_z - M_z$ is analytical in this domain, $\varepsilon = \sup(\delta/2 \colon D(\rho - \delta) = D(\rho))$ is positive, since $D(\rho)$ is finite for $\rho > \omega(P)$. The definition of $D(\rho)$ easily yields that Λ_z is analytical in view of assertion (b) of Theorem 4.1 and the aperiodicity of X, since, by Theorem 2.2, the spectrum of P on the unit circle contains only the unity and the corresponding projector is equal to Π.

The first summands on the right-hand sides of (4.9) and (4.10) coincide, and the norm of the second one in (4.10) (that equals to $O_\rho(t)$) is estimated by $O\left((\rho - \varepsilon)^t\right)$, since the function Λ_z is analytical and bounded for $|z| > \rho - \varepsilon$. Then the theorem is proved. □

4.2. Expansions in stability theorems

We consider in this paragraph a chain X along with another Markov chain $Y = (Y_t, t \geq 0)$ with the transition kernel $Q(x, A)$ represented in the form $Q = P + D$. The perturbation kernel $D \in \mathcal{B}$ is supposed to be small in the operator norm (1.3). It is also assumed that the norm $\|\cdot\|$ satisfies the conditions (M1)–(M3).

THEOREM 4.3. *Let a chain X be uniformly ergodic in a norm $\|\cdot\|$. Then*

(a) *any stochastic kernel $Q = P + D$ has, for sufficiently small $\|D\|$, a unique invariant stochastic measure ν and for all $n \geq 1$,*

$$\|\nu - \pi\pi DR - \cdots - \pi(DR)^{n-1}\| = O\left(\|D\|^n\right) \tag{4.11}$$

as $\|D\| \to 0$, where R is the generalized potential (1.8);

(b) *expansion* (4.11) *is uniform on the class of all Markov chains X with bounded norms of the potentials R and of invariant measures π.*

PROOF. The proof of Theorem 4.3 immediately follows from the boundedness of the potential R and the projector Π (see, Theorem 1.3), and also from the stability estimate for X in Theorem 5.3.

Asymptotical expansions for the finite-time distributions are expressed through the sequence of operators

$$\begin{gathered} U_t^{(n)}(D) = \sum_{s_1 + \cdots + s_{n+1} = t-n, s_i \geq 0} P^{s_1} D\left(P^{s_2} - \Pi\right) \cdots D\left(P^{s_{n+1}} - \Pi\right), \\ R_t^{(n)}(D) = \sum_{k=1}^{n} U_t^{(k)}(D), \end{gathered} \tag{4.12}$$

where $\Pi = 1 \circ \pi$ is the stationary projector of the chain X and, by the definition, $U_t^{(n)}(D) = 0$ for $t < n$, $R_t^{(0)}(D) = 0$. Note that $\|U_t^{(n)}(D)\| = O\left(\|D\|^n\right)$ as $\|D\| \to 0$ and, for $D = \delta C$, $U_t^{(n)}(\delta C) = \delta^n U_t^{(n)}(C)$. Hence $R_t^{(n)}(D)$ is an operator polynomial of D with a degree that does not exceed n. □

THEOREM 4.4. *Let condition* (C) *be satisfied and a chain X be uniformly ergodic in a norm $\|\cdot\|$ and aperiodic. Then*

(a) *for any chain Y with the kernel $Q = P + D$ and for all $k \geq 0$*

$$\sup_{ty \geq 1} \left| \mathsf{E}_\alpha f(Y_t) - \mathsf{E}_\alpha f(X_t) - \alpha R_t^{(k)}(D) f \right| = O\left(\|D\|^{k+1}\right) \tag{4.13}$$

as $\|D\| \to 0$ *uniformly in* α *and* f *with* $\|\alpha\| \le c$, $\|f\| \le c$;

(b) *expansion* (4.13) *is uniform on the class of all Markov chains* X *with bounded norm of the projector* Π *and satisfying condition* (B) *or condition* (T) *for a fixed* ρ *and bounded* n *and* m.

PROOF. Define operators

$$D_t = Q^t - P^t, \qquad B_t = D(P^t - \Pi), \qquad D_t^{(k)} = D_t - R_t^{(k)}. \tag{4.14}$$

Note that the expression under the modulus sign in (4.13) coincides with $\alpha D_t^{(k)} f$. Thus, to prove (4.13) it is sufficient to estimate the right-hand side of the inequality

$$\left|\alpha D_t^{(k)} f\right| \le \|\alpha\| \cdot \|D_t^{(k)}\| \cdot \|f\|. \tag{4.15}$$

LEMMA 4.1. *For all* $k \ge 0$ *and* $t \ge 0$,

$$D_t^{(k)} = U_t^{(k+1)}(D) + \sum_{s=0}^{t-1} D_s^{(k)} B_{t-1-s}. \tag{4.16}$$

PROOF. The proof of Lemma 4.1 is based on induction. Let $k = 0$. Then, for $t = 0$, the left and right-hand sides of (4.16) are equal to zero, since $D_0^{(0)} = D_0 = Q^0 - P^0 = 0$ and $U_0^{(1)}(D) = 0$ by definition. Now we represent the operator $D_t^{(0)} = D_t = Q^t - P^t$ as

$$D_t^{(0)} = Q^{(t-1)}(Q-P) + Q^{(t-2)}(Q-P)P + \cdots + (Q-P)P^{t-1}.$$

Recalling the definition of $D = Q - P$ and the identity $D\Pi = 0$ that follows from the equality $\Pi = 1 \circ \pi$, we obtain

$$\begin{aligned} D_t^{(0)} &= \sum_{s=0}^{t-1} Q^s D\left(P^{t-1-s} - \Pi\right) = \sum_{s=0}^{t-1} Q^s B_{t-1-s} \\ &= \sum_{s=0}^{t-1} P^s B_{t-1-s} + \sum_{s=0}^{t-1} D_s^{(0)} B_{t-1-s} = U_t^{(1)}(D) + \sum_{s=0}^{t-1} D_s^{(0)} B_{t-1-s}. \end{aligned}$$

So (4.16) is valid for $k = 0$ and all $t \ge 0$.

Let (4.16) hold true for some k. Note that, according to (4.14) and (4.12), $D_t^{(k+1} = D_t^{(k)} - U_t^{(k+1)}(D)$. Furthermore, it follows from (4.14) and (4.12) that

$$\sum_{s=0}^{t-1} U_s^{(k+1)}(D) B_{t-1-s} = U_t^{(k+2)}(D). \tag{4.17}$$

Substituting $D_t^{(k)} = D_t^{(k+1)} + U_t^{(k+1)}(D)$ into (4.16) and using (4.17) we conclude that

$$D_t^{(k+1)} = \sum_{s=0}^{t-1} \left(D_s^{(k+1)} + U_s^{(k+1)}(D)\right) B_{t-1-s} = U_t^{(k+2)} + \sum_{s=0}^{t-1} D_s^{(k+1)} B_{t-1-s}.$$

Thus identity (4.16) holds true also for $k+1$. The lemma is proved. □

To complete the proof of the theorem observe that the sum $S(P) = \sum \|P^t - \Pi\|$ is bounded and

$$\sum \|B_s\| \le \|D\| S(P) < \frac{1}{2} \tag{4.18}$$

for $\|D\| < 1/2S(P)$. Therefore, for all such D, from (4.16) we conclude that

$$\sup_{t\ge 0} \|D_t^{(k)}\| \le \sup_{t\ge 0} \|U_t^{(k+1)}(D)\| + \sum_{s\ge 0} \|B_s\| \sup_{t\ge 0} \|D_t^{(k)}\|.$$

Hence by (4.18)

$$\sup_{t\ge 0} \left\|D_t^{(k)}\right\| \le 2 \sup_{t\ge 0} \left\|U_t^{(k+1)}(D)\right\| \tag{4.19}$$

as $\|D\| < 1/2S(P)$. Since $\sup \|P^t\| \le \|\Pi\| + S(P)$, it follows from the definition (4.12) that

$$\begin{aligned} \sup_{t\ge 0} \|U_t^{(k+1)}(D)\| &\le \sup_{t\ge 0} \|P^t\| \Big(\sum_{s\ge 0} \|D(P^s - \Pi)\| \Big)^{k+1} \\ &\le (\|\Pi\| + S(P))\,(S(P))^{k+1}\,\|D\|^{k+1}. \end{aligned} \tag{4.20}$$

Substituting (4.20) into (4.19) and using (4.15) we make sure that assertion (a) of the theorem is valid.

Assertion (b) immediately follows from estimates (4.15), (4.19), and (4.20) and from theorems 3.4 and 3.7, since the sum $S(P)$ is uniformly bounded in the class of chains under consideration. The theorem is proved. □

4.3. Applications of asymptotic expansions

Consider some examples of applications of representation (4.9) where the index $\omega(P)$, the sets $D(\rho)$, and the operators $C_k(\xi)$ may be implicitly evaluated. We examine in this paragraph a countable Markov chain X with values in $E = \mathbb{Z}_+$ given by its transition matrix $P = (p_{ij}, i, j \in E)$. Denote by $\pi = (\pi_j, j \in E)$ the invariant stochastic measure of the chain, and by (p_{ij}^t) the transition probabilities over t steps.

We take $\mathfrak{M}_v$, that is, the space of all measures $\mu = (\mu_j, j \ge 0)$ with a finite norm $\|\mu\| = \sum |\mu_j| v^j$, as the space $\mathfrak{M}$, where $v \ge 1$ is a fixed real parameter. The corresponding operator norm (1.5) of a matrix $Q = (q_{ij})$ is of the form

$$\|Q\|_v = \sup_{i\ge 0} \sum_{j\ge 0} |q_{ij}| v^{j-i}.$$

Denote by $\omega(v)$ the index (4.1) calculated for the transition matrix P and the norm $\|\cdot\|$.

LEMMA 4.2. *Let* $\|P\|_v < \infty$. *Then*

$$\omega(v) = \inf_{n\geq 1} \liminf_{N\to\infty} (\|P_N^n\|_v)^{1/n}, \tag{4.21}$$

where the matrix P_N^n *is obtained from* P^n *by replacing the first* N *rows by zero rows.*

PROOF. Choosing in (4.1) a finite-dimensional operator $P^n - P_N^n$ for K, we make sure that $\omega(v)$ does not exceed the right-hand side of (4.21).

To obtain the converse inequality, note that we may evaluate the lower bound in (4.1) (for $E = \mathbb{Z}_+$) only for all matrices K with finite number of nonzero rows, since each compact operator is approximated by finite-dimensional operators. Furthermore, the norm of the matrix $P^n - K$ reduces after replacing their rows by zero rows. Then for all such K, $\|P^n - K\|_v \geq \|(P^n - K)_N\|_v = \|P_N^n\|_v$ for all sufficiently big N, whence (4.21) follows. The lemma is proved. □

EXAMPLE 4.2 (LOWER CONTINUOUS RANDOM WALK). Let $p_{ij} = g_{j-i+1}$ for $j \geq i-1$, $p_{0j} = g_j$, where $g = (g_j, j \geq 0)$ is a discrete probability distribution on E. Denote by $g(z) = \sum z^n g_n$ the generating function of g and assume that a chain X is exponentially ergodic, i.e.,

$$m = \sum n g_n < 1, \qquad g(v) < \infty, \tag{4.22}$$

for some $v > 1$. It follows from (4.22) that the index

$$w = \sup(v \geq 1 : g(v) \leq v) \tag{4.23}$$

is strictly greater than the unity. Furthermore, $g(v) < v$ if and only if $1 < v < w$.

LEMMA 4.3. *The following equality holds true*

$$\omega(v) = \frac{g(v)}{v} \tag{4.24}$$

so the chain X *is uniformly ergodic if and only if* $1 < v < w$. *The equality* (4.24) *remains valid also after changing a finite number of rows of the matrix* P.

PROOF. Use (4.21) and note that for $N > n$

$$\|P_N^n\|_v = \sup_{i>N} \sum_{j\geq 0} p_{ij}^n v^{j-i} = \left(\frac{g(v)}{v}\right)^n,$$

since the stochastic sequence $X_0 = i, X_1, \ldots, X_n$ coincides in the domain $i > N$ with a random walk without boundaries. □

LEMMA 4.4. *Let $1 < v < w$. Then for any $\rho \in (\omega(v), 1)$ the set $D(\rho)$ is empty.*

PROOF. In view of Remark 4.2, it is sufficient to prove that the operator $\xi - P: \mathfrak{M}_v \to \mathfrak{M}_v$ has no annihilators for $\omega(v) < |\xi| < 1$.

Let $\xi\mu = \mu P$ for a measure $\mu = (\mu_j) \in \mathfrak{M}_v$ and $|\xi| > \omega(v)$, $\xi \neq 1$. Then $(\xi z - g(z))\mu(z) = \mu_0(z-1)g(z)$, where the generating function $\mu(z) = \sum \mu_j z^j$ of the measure μ is analytical in the domain $|z| < v$ and is continuous on its boundary. Since $|g(z)| \leq g(v) = v\omega(v) < |\xi z|$ for $|z| = v$, by the Rouchet theorem [see Schwartz (1967), Theorem 68], the equation $\xi z = g(z)$ has a unique solution $z = z_\xi$ in the domain $|z| < v$. It is clear that $g(z_\xi) = \xi z_\xi \neq 0$ and $z_\xi \neq 1$, since $\xi \notin \{0, 1\}$. Therefore from the equation above for $\mu(z)$ and from the analyticity of $\mu(z)$ we conclude that $\mu_0 = 0$ for $z = z_\xi$, whence $\mu(z) = 0$. The lemma is proved. □

THEOREM 4.5. *Let the distribution g of jumps of a chain X satisfy conditions* (4.22). *Then for any $1 < v < w$ and for all $\varepsilon > 0$ the asymptotic expansion is valid*

$$\sup_{i \geq 0} \sum_{j \geq 0} \left| p_{ij}^t - \pi_j \right| v^{j-1} = O\left((\omega(v) + \varepsilon)^t \right), \tag{4.25}$$

where w and $\omega(v)$ are calculated by (4.23) *and* (4.24) *respectively.*

The proof of the theorem immediately follows from Lemmas 4.3 and 4.4 and from expansion (4.9) of Theorem 4.2. The aperiodicity of X follows from (4.22), since $g_0 > 0$.

COROLLARY 4.1. *Under conditions of Theorem* 4.3 *the following inequality is valid*

$$\sup_{i \geq 0} \limsup_{n \to \infty} \left(\left| p_{ij}^n - \pi_i \right| \right)^{1/n} \leq \inf_{v \geq 1} \frac{g(v)}{v}. \tag{4.26}$$

Note that (4.26) is an equality for a Bernoulli walk on E, i.e., for a chain X with $g_2 = 1 - g_0 < g_0$. In this case, the left-hand and right-hand sides of (4.26) are equal to $2(g_0 - g_0^2)^{1/2}$. The last index is the greatest by absolute value point of the spectrum of the operator P, excluding unity.

The proof of Corollary 4.1 is obtained from (4.25), since the sum on the left-hand side of (4.25) is not less than $|p_{ii}^t - \pi_i|$ that does not depend on v and the right-hand side does not exceed $c(v)(\omega(v) + \varepsilon)^t$ where $c(v)$ does not depend on i.

EXAMPLE 4.3 (UPPER CONTINUOUS RANDOM WALK). Let $p_{ij} = g_{i+1-j}$ for $1 \leq j \leq i+1$, $i > 0$, $p_{01} = 1$, and $p_{i0} = G_i = \sum_{j > i} g_j$, where $g = (g_j, j \geq 0)$ is a discrete probability distribution on $E = \mathbb{Z}_+$ with the generating function $g(z)$. Assume that a chain X is ergodic and aperiodic, i.e.,

$$m = \sum n g_n > 1, \qquad g_1 G_1 > 0. \tag{4.27}$$

Then, for $g_0 > 0$, the equation

$$wg\left(\frac{1}{w}\right) = 1 \tag{4.28}$$

has a unique solution $w > 1$. Put also $w = \infty$ for $g_0 = 0$. From the convexity of $g(z)$ we conclude that the inequality $vg(1/v) < 1$ is valid if and only if $1 < v < w$.

LEMMA 4.5. *The equation*

$$\omega(v) = vg\left(\frac{1}{v}\right) \tag{4.29}$$

is valid, so the chain X is uniformly ergodic in a norm $\|\cdot\|_v$ if and only if $1 < v < w$.

The proof of Lemma 4.5 immediately follows from (4.21), since

$$\|P_N^n\|_v = \sup_{i>N} \sum_{j\geq 0} p_{ij}^n v^{j-i} = \lim_{i\to\infty} \sum_{j\geq 0} p_{ij}^n v^{j-i} = \left(vg\left(\frac{1}{v}\right)\right)^n.$$

Note that the function $\omega(v)$ decreases to its minimum on the interval $(1, w)$ and then increases, since the derivative

$$\omega'(v) = g\left(\frac{1}{v}\right) - \frac{g'\left(\frac{1}{v}\right)}{v} = g_0 - \frac{g_2}{v^2} - \frac{2g_3}{v^3} - \cdots$$

changes sign at most one time, and $\omega'(1) = 1 - m < 0$, $\omega(1) = 1 \geq \omega(w)$.

Denote by u the point of the minimum of $\omega(v)$ and by $r < 1$ the corresponding minimal value:

$$r = \min_{s\geq 1} sg\left(\frac{1}{s}\right) = ug\left(\frac{1}{u}\right). \tag{4.30}$$

If $g_0 = 0$ then $r = g_1 < 1$ and $u = \infty$. Let also

$$h = \frac{g_0}{1 - g_0}, \qquad \theta = \frac{g(-h)}{h}. \tag{4.31}$$

The definition of r and the convexity of g imply that, for $\theta \in (r, 1)$, the equation $\omega(v) = \theta$ has two roots v_1, v_2 such that $1 < v_1 < u < v_2 < w$ for $g_0 > 0$ and $1 < v_1 < u < v_2 = w = \infty$ for $g_0 = 0$. Furthermore, $\omega(v) < \theta$ if and only if $v_1 < v < v_2$.

LEMMA 4.6. *Let $1 < v < w$.*

(a) *If $uh > 1$ or $g(-h) \notin (rh, h)$ then the set $D(\rho)$ is empty for all $\rho \in (\omega(v), 1)$;*
(b) *If $uh < 1$ and $g(-h) \in (rh, h)$ then $D(\rho) = \{-\theta\}$ for $\rho \in (\omega(v), \theta]$ and $D(\rho) = \varnothing$ for other $\rho > \omega(v)$.*

REMARK 4.3. The condition (b) of the lemma does not depend on v. If this condition is satisfied then the spectral set $D(\omega)$ in (4.7) constructed by the norm $\|\cdot\|_v$ is nonempty for some $v \in (1, w)$, namely for $v \in (v_1, v_2)$.

Condition (b) is fulfilled for small g_0. For example, for the distribution $g_j = \varepsilon(1-\varepsilon)^j$, $j \geq 0$, one can evaluate $u = 2(1-\varepsilon) < (1-\varepsilon)/\varepsilon = h^{-1}$, $rh = 4\varepsilon^2 < \varepsilon/(1+\varepsilon) = g(-h) < h$ provided that $\varepsilon < \frac{1}{4}$.

PROOF OF LEMMA 4.6. Let $D(\rho) \neq \varnothing$ for some $v \in (1, w)$, $\rho \in (\omega(v), 1)$ and $\xi \in D(\rho)$. Then, by Remark 4.2, the matrix $\xi - P$ has an annihilator in the dual space $\mathfrak{N}_v$ of functions $f = (f_i)$ with a finite norm $\|f\|_v = \sup v^{-i}|f_i|$. Thus $\xi f = Pf$ for some $f \in \mathfrak{N}_v$.

Denote by $f(z) = \sum f_i z^i$ the generating function of f (it is analytical in the circle $|z| < v^{-1}$ because of $f \in \mathfrak{N}_v$). The equation for f implies

$$(\xi z - g(z))f(z) = f_0 \left[z(1-g_0)(\xi - 1) + \frac{g(z)-z}{z-1} \right]. \tag{4.32}$$

The inequality $|\xi| \ge \rho > \omega(v)$ for $|z| = v^{-1}$ and Lemma 4.5 imply $|g(z)| \le g(v^{-1}) = \omega(v)/v = \omega(v)|z| < |\xi z|$. Therefore the equation $\xi z = g(z)$ has a unique solution $z = z_v$ in the domain $|z| < v^{-1}$. So z_v is a solution to the right-hand side of (4.32). However the system $\xi z = g(z)$, $z(1-g_0)(\xi-1) + (g(z)-z)/(z-1) = 0$ that is equivalent to the system $\xi z = g(z)$, $(\xi - 1)z(z(1-g_0) + g_0) = 0$, has a unique solution $z_v = -h$ only for $\xi = -\theta$. Therefore, $h = |z_v| < v^{-1}$ and $\theta = |\xi| \ge \rho > \omega(v)$, $\theta = |\xi| < 1$.

Now we assume that $h < v^{-1}$ and $\omega(v) < \theta < 1$. Since the solution z_v of the left-hand side of (4.32) is unique in the domain $|z| < v^{-1}$, a nontrivial solution $f(z)$ of the equation (4.32) is analytical in a neighborhood of the circle $|z| < v^{-1}$ because $|g(z)| < |\xi z|$ on its boundary and the singularity at the point $z = -h$ is removable. Therefore, a solution to (4.32) that has the form

$$f(z) = \frac{1}{1-z} \left[\frac{(z+h)z}{\theta z + g(z)} - \frac{1+h}{\theta + 1} \right] \tag{4.33}$$

is the generating function of the unique annihilator $f \in \mathfrak{N}_v$ of the operator $\xi - P$ for $\xi = -\theta$. For other ξ such annihilators do not exist.

It remains to check that the system of inequalities

$$h < v^{-1}, \qquad \omega(v) < \theta < 1, \tag{4.34}$$

has a solution $v \in (1, w)$ if and only if condition (b) holds true. Indeed, from (4.24) and monotonicity of $g(z)$ we conclude that $\theta = g(-h)/h < g(h)/h = \omega\left(h^{-1}\right)$. It follows from properties of ω that in this case $h^{-1} \in (1, v_1) \cup (v_2, \infty)$. Since $\omega(v) < \theta$, and so $v \in (v_1, v_2)$, and $h^{-1} > v$, $h^{-1} \in (v_2, \infty$, $h^{-1} > v_2 > u$, whence $hu < 1$. Thus

$$\frac{g(-h)}{h} = \theta > \omega(v) \ge \min \omega(s) = r$$

and $g(-h) < h$. Then $g(-h) \in (rh, h)$. On the other hand, under (b), from $h^{-1} > u$ and the inequality $\omega\left(h^{-1}\right) > \theta$ we get $h^{-1} > v_2 > v$ for all v with $\omega(v) < \theta$. Hence the system (4.34) is equivalent to the inclusion $v \in (v_1, v_2)$. This interval is nonempty, since $\theta = g(-h)/h > r = \min \omega(s)$. The lemma is proved. □

Since the annihilator $f \in \mathfrak{N}_v$ is unique (provided it exists), the Fredholm property of the operator $\xi - P$ for $|\xi| > \omega(P)$ implies by Theorem 4.1 (c) that the matrix $\xi - P = -\theta - P$ has a unique annihilator $\nu \in \mathfrak{M}_v$. Simple calculations show that

$$\nu = (\nu_i, i \ge 0), \qquad \nu_0 = g_0, \qquad \nu_i = (-h)^i, \quad i \ge 1, \tag{4.35}$$

and the action of ν on f is equal to

$$\nu f = \sum \nu_i f_i = f(-h) = \frac{g_0 h}{g(-h) + h g'(-h)}. \tag{4.36}$$

Here $g(-h)+hg'(-h) = g_0-(-h)^2g_2-2(-h)^3g_3-\cdots \geq g(h)-hg'(h) = \omega'\left(h^{-1}\right) > 0$, since, by Lemma 4.6, $h^{-1} > u$ provided that an annihilator exists, and the function $\omega(v)$ increases for $v > u$ by the definition of u. Therefore the projector $C_n(\xi)$ in (4.4) is equal to the matrix $\left((\nu f)^{-1}f_i\nu_j, i,j \geq 0\right)$. Thus, it follows from equations (4.8), for $n(\xi) = n > 1$, that $0 = \nu(\xi - P)C_{n-1}(\xi) = \nu = C_n(\xi) = (\nu f)\nu \neq 0$. This contradiction proves that the multiplicity $n(\xi)$ of the point $\xi = -\theta$ in the spectrum of the operator P is equal to the unity, that is, $n(\xi) = 1$. Applying (4.9) and Lemma 4.5 we arrive at the following assertion.

THEOREM 4.6. *Let a chain X satisfy conditions (4.27). Then for any $v \in (1, w)$ and all $\varepsilon > 0$ the following asymptotic expansion is valid*

$$\sup_{i\geq 0} \sum_{j\geq 0} v^{j-i} \left| p_{ij}^t - \pi_j - \delta_v(-\theta)^t \frac{f_i}{\nu f} \right| = O\left((\omega(v)+\varepsilon)^t\right) \tag{4.37}$$

as $t \to \infty$, where w and $\omega(v)$ are given by (4.28) and (4.29) respectively and θ, h, f_i, ν_j, and νf are determined in (4.31), (4.33), (4.35), and (4.36) respectively. The indicator δ_v is equal to 1, for $\omega(v) < \theta < 1$ and $v < h^{-1}$, and to 0 otherwise.

COROLLARY 4.2. *Under the conditions of Theorem 4.4, the following inequality*

$$\sup_{i\geq 0} \limsup_{t\to\infty} \left(\left| p_{ii}^t - \pi_i - \delta(-\theta)^t f_i \frac{\nu_i}{\nu f} \right| \right)^{1/t} \leq r, \tag{4.38}$$

holds true, where $r = \min g(s)/s$, $\delta = 1$ under the conditions (b) *of Lemma 4.6 and $\delta = 0$ otherwise.*

The proof of (4.38) is based on (4.37) and is the same as that of Corollary 4.1.

CHAPTER V

STRONG STABILITY AND INDUCED NORM OF PERTURBATION

In this Chapter, we investigate the stability of distributions for a Markov chain $X = (X_t, t \geq 0)$ with the kernel $P(x, A)$ without an assumption of the uniform ergodicity for this chain.

It was showed in Chapter 3 that the stability of the invariant distribution of a perturbed chain $Y = (Y_t, t \geq 0)$ with the kernel $Q = P + D$ does not follow from the operator convergence to zero of a perturbation D. We define below, on a class of perturbation kernels D, an induced norm (by the basic norm and by the kernel P) such that the convergence $D \to 0$ implies the strong stability.

In order that this norm is finite, it is necessary and sufficient that the perturbation D be subordinated to the potential (generally speaking, unbounded) R of the basic chain. Under this assumption we obtain an asymptotic expansion for the perturbed invariant measure with explicit bounds of the remainder term. The finite-time distributions are also considered. The main unbounded part of the generalized potential R is described by means of distributions of hitting times of a positive set.

We suppose in the sequel that the chain X has a unique invariant stochastic measure π and satisfies condition (P) of Chapter 2 and condition (C) of Chapter 1. Conditions (M1)-(M3) of Chapter 1 are supposed to be satisfied for a norm on $\mathfrak{M}$. Condition (C) yields that there exists a set $C \in \mathcal{E}$ such that $\pi(C) > 0$ and $r = \alpha(C) \inf_{x \in C} h(x) > 0$. Furthermore, (C) implies the following minorization condition

$$P^n(x, A) \geq r\beta(A) \tag{5.1}$$

for all $x \in C$, $A \in \mathcal{E}$, where the probability distribution $\beta(A) = \alpha(A \cap C)/\alpha(C)$ is concentrated on C and n, α, h are involved in condition (C).

5.1. Stability of the invariant distribution

Denote by $\Pi = \mathbb{I} \circ \pi$ the stationary projector of the chain X and consider the generalized potential R and the subspace $\mathfrak{M}_0 = \mathfrak{M}(I - P + \Pi)$ that are defined in (1.8) and (1.9).

Define also the set of invariant measures in $\mathfrak{M}$ for the kernel Q as follows

$$\mathcal{N}(Q) = \{\nu \in \mathfrak{M}^+ : \nu = \nu Q,\ \nu\mathbb{I} = 1\}.$$

Recall that by Lemma 2.1 the linear hull of $\mathcal{N}(Q)$ coincides with the set of all invariant measures of Q.

THEOREM 5.1. *If $\mathcal{N}(Q)$ is nonempty then the inclusion $\nu \in \mathcal{N}(Q)$ is valid if and only if $\nu D \in \mathfrak{M}_0$ and*

$$\nu - \pi = \nu D R. \tag{5.2}$$

Under additional assumption $\pi D \in \mathfrak{M}_0$ the inclusion $\nu \in \mathcal{N}(Q)$ is equivalent to $\nu D \in \mathfrak{M}_0$, $\nu DRD \in \mathfrak{M}_0$, and

$$\nu - \pi - \pi D R = \nu D R D R. \tag{5.3}$$

PROOF. Obviously, $\nu = \nu Q$ if and only if $\nu(I-P) = \nu - \nu P = \nu Q - \nu P = \nu D$. Note that $\pi = \pi P$, $D\Pi = (D\mathbb{1})\pi = 0$, since Q is a stochastic kernel, and $(\nu-\pi)\Pi = 0$. By Definition 1.3 of the potential R, $(\nu-\pi)(I-P) = \nu(I-\Pi) = \nu D = \nu D(I-\Pi)$ is equivalent to (5.2). If $\pi D \in \mathfrak{M}_0$ then $\pi DR \in \mathfrak{M}$ and from (5.2) we conclude that $\nu DRD = (\nu-\pi)D = \nu D - \pi D \in \mathfrak{M}_0$. Thus (5.2) is equivalent to equations

$$\begin{aligned}(\nu - \pi - \pi DR)(I-P) &= (\nu-\pi)(I-P) - \pi D(I-\Pi)\\ &= \nu D - \pi D = \nu DRD = \nu DRD(I-\Pi).\end{aligned}$$

The latter is equivalent to (5.3) by Definition 1.3. The theorem is proved. □

Theorem 5.1 implies that under a stability effect, i.e. if $\mathcal{N}(Q) \to \{\pi\}$ as $D \to 0$, the set of invariant measures $\nu \in \mathcal{N}(Q)$ is transformed by the perturbation D from a neighborhood of the measure π into the space $\mathfrak{M}_0$. This set of measures ν may contain the whole neighborhood of π. This is the case, for example, if the perturbation $D = \varepsilon B$ is analytical, where $B \in \mathcal{B}$ is fixed. Therefore, to be sure of the stability under perturbation D, it is reasonable to assume that the image of the operator D is contained in the subspace $\mathfrak{M}_0$. An exact statement is given in the following result.

THEOREM 5.2. *Let a linear subspace $\mathfrak{M}_1 \subset \mathfrak{M}$ be such that $\mathfrak{M}_0 \subset \mathfrak{M}_1$ and $\mathfrak{M}_1 \setminus \mathfrak{M}_0$ contain at least one measure β satisfying inequality* (5.1) *for all $x \in C$, $A \in \mathcal{E}$, and for some $C \in \mathcal{E}$, $\pi(C) > 0$, $r > 0$, $n \geq 1$. Then there exists a set of stochastic kernels $Q_\varepsilon = P + D_\varepsilon$ such that*

(a) $\mathfrak{M}D_\varepsilon \subset \mathfrak{M}_1$,

(b) $\|D_\varepsilon\| \to 0$ *as* $\varepsilon \to 0$ *and either* $\mathcal{N}(Q_\varepsilon) = \varnothing$ *for all* $\varepsilon > 0$ *or*

$$\inf_{\varepsilon>0} \inf\left(\|\nu - \pi\|, \nu \in \mathcal{N}(Q_\varepsilon)\right) > 0.$$

PROOF. We choose a measure $\beta \in \mathfrak{M}_1 \setminus \mathfrak{M}_0$ in accordance with (5.1) so that $\beta(C) = 1$. Then $P^{n+1} \geq rh \circ \beta$ with the function $h(x) = P(x, C)$ and the kernel

$$Q_\varepsilon = (1-\varepsilon)P + \varepsilon P^{n+1} + \varepsilon^2 h \circ (\pi - \beta) \tag{5.4}$$

is stochastic for $\varepsilon \leq r$. Let $P_n = \sum_{s=0}^{n-1}(P^s - \Pi)$. It is easily seen that the measure

$$\mu D_\varepsilon = \mu(Q_\varepsilon - P) = -\varepsilon\mu P P_n(I - P + \Pi) + \varepsilon^2(\mu h)\pi - \varepsilon^2(\mu h)\beta, \tag{5.5}$$

belongs to the space $\mathfrak{M}_1$ for all $\mu \in \mathfrak{M}$, since all summands in (5.5) except the last one are contained in $\mathfrak{M}_0 \subset \mathfrak{M}_1$ and $\beta \in \mathfrak{M}_1$. Besides, $\|D_\varepsilon\| = O(\varepsilon)$ as $\varepsilon \to 0$.

If $\mathcal{N}(Q_\varepsilon) \neq \varnothing$ for an infinite sequence of indices $\varepsilon \to 0$ then substituting $\mu = \nu_\varepsilon \in \mathcal{N}(Q_\varepsilon)$ into (5.5) we obtain by Theorem 5.1 that $\nu_\varepsilon D_\varepsilon = \nu_\varepsilon(Q_\varepsilon - P) \in \mathfrak{M}_0$ and $\nu_\varepsilon h = 0$ because of $\beta \notin \mathfrak{M}_0$. The equality $\nu_\varepsilon h = 0$ that holds for an infinite sequence of $\varepsilon \to 0$ contradicts the convergence $\nu_\varepsilon \to \pi$ in the norm $\|\cdot\|$, since the latter yields the convergence in the total variation. On the other hand $\pi h = \pi(C) > 0$. The theorem is proved. □

In view of Theorems 5.1 and 5.2 we assume in the sequel that $\mathfrak{M}D \subset \mathfrak{M}_0$. In this case the operator norm of the product DR is a natural index for the deviation $\nu - \pi$ in (5.2). We shall use the following auxiliary result.

LEMMA 5.1. *Let $\mathfrak{M}D \subset \mathfrak{M}_0$. Then the operator DR is well defined on $\mathfrak{M}$ and is bounded.*

PROOF. The proof of the lemma immediately follows from Theorem 1.2 that implies $\mu DR \in \mathfrak{M}$ provided that $\mu D \in \mathfrak{M}_0$. The last inclusion follows from $\mu \in \mathfrak{M}$. The boundedness of DR arises from the closed graph theorem, [see Kato (1966), Theorem 3.5.20]. Indeed, if $\mu_n = \nu_n DR$ then by Definition 1.2 $\mu_n \mathbb{1} = 0$, $\mu_n(I-P) = \nu_n D(I-\Pi)$ and by $\mu_n \to \mu$, $\nu_n \to \nu$, $\mu\mathbb{1} = 0$, $\mu(I-P) = \nu D(I-\Pi)$ because P and D are bounded. The last relations are equivalent to $\mu = \nu DR$ by the definition of a potential. The lemma is proved. □

THEOREM 5.3. *Let $\mathfrak{M}D \subset \mathfrak{M}_0$ and $\varepsilon = \|DR\| < 1$. Then there exists a unique invariant stochastic measure $\nu \in \mathfrak{M}$ of the kernel Q and for all $n \geq 1$ the following estimates*

$$\left\|\nu - \pi - \pi DR - \cdots - \pi(DR)^{n-1}\right\| \leq \|\nu\|\varepsilon^n \leq \|\pi\|\frac{\varepsilon^n}{1-\varepsilon} \tag{5.6}$$

are valid, where π is the invariant measure and R is the generalized potential for the basic chain X defined by the unperturbed kernel P.

REMARK 5.1. If a chain X is strongly stable (i.e., uniformly ergodic) in the norm $\|\cdot\|$ then by Theorem 1.3 $\mathfrak{M}_0 = \mathfrak{M}$, the inclusion $\mathfrak{M}D \subset \mathfrak{M}_0$ holds true for all bounded perturbations $D \in \mathcal{B}$, and $\varepsilon = \|DR\| = O(\|D\|)$ as $\|D\| \to 0$, since the potential R is bounded.

PROOF. Let $\nu \in \mathcal{N}(Q)$. Then (5.2) gives us $\nu(I - DR) = \pi$ where the operator $I - DR$ has a bounded inverse, since $\|DR\| < 1$. Thus equation (5.2) has a unique solution

$$\nu = \pi(I - DR)^{-1} \tag{5.7}$$

that may be represented as the Neumann series $\nu = \sum \pi(DR)^n \in \mathfrak{M}$. This representation yields that the expression under the norm sign on the left-hand side of (5.6) is equal to

$$\sum_{t \geq n} \pi(DR)^t = \nu(DR)^n,$$

whence the first inequality in (5.6) follows immediately. The second one arises from (5.7) and from the estimate $\|\nu\| \leq \sum \|\pi\| \cdot \|DR\|^n$.

On the other hand, the measure ν defined in (5.7) satisfies the relation (5.2) that is equivalent by Definition 1.2 to the system

$$(\nu - \pi)\mathbb{1} = 0, \qquad (\nu - \pi)(I - P) = \nu D(I - \Pi),$$

whence $\nu\mathbb{1} = 1$, $\nu(I - P) = \nu D = \nu(Q - P)$, and $\nu = \nu Q$. So ν is invariant for Q and $\nu\mathbb{1} = 1$. Lemma 2.1 and the uniqueness of ν gives us the nonnegativity of ν. The theorem is proved. □

Theorem 5.2 shows that the condition $\mathfrak{M}D \subset \mathfrak{M}_0$ is essential for the stability of the invariant distribution π in the norm $\|\cdot\|$. But this condition itself does not imply the convergence $\nu \to \pi$ even if $\|D\| \to 0$. Estimate (5.6) for $n = 1$ yields that it is sufficient to assume the convergence to zero of the induced operator norm $\|DR\|$ of the perturbation to obtain the stability of ν. This norm is equivalent to $\|D\|$ only for uniformly ergodic chains. The following theorem establishes the necessity of the convergence $\|DR\| \to 0$ provided that all other conditions are fulfilled.

THEOREM 5.4. *Let $\mathfrak{M}_0 \neq \mathfrak{M}$, so the potential R is unbounded. Assume that the space $\mathfrak{M}_0$ contains at least one measure β satisfying the minorization condition* (5.1). *Then, for all $\delta > 0$, there exists a family of stochastic kernels $Q_\varepsilon = P + D_\varepsilon$ such that*

(a) $\mathfrak{M}D_\varepsilon \subset \mathfrak{M}_0$ *for all* $\varepsilon > 0$,
(b) $\|D_\varepsilon\| \to 0$ *as* $\varepsilon \to 0$,
(c) $\delta/2 \le \|D_\varepsilon R\| \le 2\delta$ *for all* $\varepsilon > 0$,

and there is no stability of the invariant distribution, that is, either $\mathcal{N}(Q_\varepsilon) = \varnothing$ for all $\varepsilon > 0$ or

$$\inf_{\varepsilon>0} \inf \left(\|\nu_\varepsilon - \pi\|, \nu_\varepsilon \in \mathcal{N}(Q_\varepsilon)\right) > 0.$$

PROOF. Since R is unbounded, there exists a set of nonnegative measures $\beta_\varepsilon \in \mathfrak{M}_0$ such that $\|\beta_\varepsilon\| = o(\varepsilon)$ as $\varepsilon \to 0$ and $\|\beta_\varepsilon R\| = \delta_0 > 0$ for all $\varepsilon > 0$.

Consider a family of kernels

$$Q_\varepsilon = (1-\varepsilon)P + \varepsilon P^n + h \circ \left((\varepsilon r - \beta_\varepsilon \mathbb{I})\pi + \beta_\varepsilon - \varepsilon r \beta\right), \tag{5.8}$$

where n, $r > 0$, and the measure β are taken to satisfy condition (5.1) and the $h(x) = \mathbb{I}_{\{x \in C\}}$. It follows from (5.1) that $P^n \ge rh \circ \beta$ and (M1) implies $\beta_\varepsilon \mathbb{I} = O(\|\beta_\varepsilon\|) = o(\varepsilon)$ as $\varepsilon \to 0$. Thus, the kernel Q_ε is nonnegative for small ε. Moreover, $Q_\varepsilon \mathbb{I} = (1-\varepsilon)\mathbb{I} + \varepsilon \mathbb{I} + h(\varepsilon r - \beta_\varepsilon \mathbb{I} + \beta_\varepsilon \mathbb{I} - \varepsilon r) = 1$, since $\beta \mathbb{I} = 1$. Thus, Q_ε is a stochastic kernel.

Recall that $\beta \in \mathfrak{M}_0$. Let $\mu \in \mathfrak{M}$. Then

$$\begin{aligned} \mu D_\varepsilon &= \mu(Q_\varepsilon - P) \\ &= -\varepsilon\mu\left(P + \cdots + P^{n-1}\right)(I - P) + (\mu h)\left((\varepsilon r - \beta_\varepsilon 1)\pi + \beta_\varepsilon - \varepsilon r \beta\right). \end{aligned} \tag{5.9}$$

It is clear that the first summand on the right-hand side of (5.9) belongs to $\mathfrak{M}_0$, $\pi \in \mathfrak{M}_0$ by the definition, and $\beta_\varepsilon \in \mathfrak{M}_0$ by the construction. So, $\mu D_\varepsilon \in \mathfrak{M}_0$ and assertion (a) is valid.

Simple calculations that follow from the boundedness of P, properties of β_ε, and (5.9), show that $\|\mu D_\varepsilon\| \le a\varepsilon\|\mu\|$. Thus $\|D_\varepsilon\| = O(\varepsilon)$ as $\varepsilon \to 0$.

Finally, let us estimate the induced norm of D_ε using the identities $(I - P)R = I - \Pi$, $\pi R = 0$, and

$$\begin{aligned} \|\mu D_\varepsilon R\| &= \left\|-\varepsilon\mu\left(P + \cdots + P^{n-1}\right)(I - \Pi) + (\mu h)(\beta_\varepsilon R - \varepsilon r \beta R)\right\| \\ &\ge \|(\mu h)\beta_\varepsilon R\| - \varepsilon\left(\left\|\mu\left(P + \cdots + P^{n-1}\right)(I - \Pi)\right\| + |\mu h| r \|\beta R\|\right) \\ &\ge |\mu h|\delta_0 - b\varepsilon\|\mu\|. \end{aligned} \tag{5.10}$$

In the same manner as that which led to (5.10) we prove the upper bound $\|\mu D_\varepsilon R\| \le |\mu h|\delta_0 + b\varepsilon\|\mu\|$. Substituting $\delta_0 = \delta/\|h\|$, we make sure of assertion (b) for all sufficiently small ε.

To complete the proof of the theorem, suppose that $\nu_\varepsilon \in \mathcal{N}(Q_\varepsilon)$ for an infinite set of indices ε. According to Theorem 5.1 and the estimate (5.10), for $\mu = \mu_\varepsilon$,

$$\|\nu_\varepsilon - \pi\| = \|\nu_\varepsilon D_\varepsilon R\| \ge (\nu_\varepsilon h)\delta_0 - b\varepsilon\|\nu_\varepsilon\|.$$

If $\nu_\varepsilon \to \pi$ in the norm $\|\cdot\|$ were the case, $\nu_\varepsilon h$ would be equal to $\nu_\varepsilon(C) \to \pi(C) = \pi h > 0$ by conditions (M1) and (5.1). So the limit of the right-hand side of the last inequality as $\varepsilon \to 0$ must be positive. This contradicts the convergence of the left-hand side to zero. Thus, $\liminf \|\nu_\varepsilon - \pi\| > 0$ and the theorem is proved. □

5.2. Main part of a potential operator

The problem of estimation of the index $\varepsilon = \|DR\|$ in Theorem 5.3 and description of the space $\mathfrak{M}_0$ can be reduced to a slightly simpler problem of investigation of a nonnegative kernel corresponding to the chain X. This kernel determines mean hitting numbers of measurable sets until the first achievement of a set $C \in \mathcal{E}$ involved in condition (5.1). Additionally to (5.1), assume that the chain X is aperiodic in the sense of Definition 2.1.

Define a Markov moment

$$\tau_C = \inf(t \geq 1 \colon X_{nt} \in C) \tag{5.11}$$

and a nonnegative kernel

$$R_C(x, A) = \sum_{t \geq 0} P_x(\tau_C > t, X_{nt} \in A). \tag{5.12}$$

Observe that the action $R_C f$ of the kernel R_C is well defined for a nonnegative measurable function f as the monotone pointwise limit of actions of partial sums of the series (5.12) on f. If we assume that the function $\mathsf{E}_x \tau_C = R_C(x, E)$ is finite for all $x \in E$ then, as shown above, the measure μR_C is σ-finite and σ-additive for any finite measure μ. Therefore, under conditions of Theorem 5.5, the following subspaces are well defined

$$\begin{aligned} \mathfrak{M}_C &= \{\mu \in \mathfrak{M} \colon \mu(I - \Pi)R_C \in \mathfrak{M}\}, \\ \mathfrak{M}_{Cn} &= \{\mu \in \mathfrak{M} \colon \mu P_n \in \mathfrak{M}_C\}, \end{aligned} \tag{5.13}$$

where $P_n = \sum_{s=0}^{n-1} (P^s - \Pi)$.

THEOREM 5.5. *Let the following condition be satisfied.*

(B) *There exists a measurable nonnegative finite function φ on E such that $\sup_{x \in C} \varphi(x) = M_\varphi < \infty$ and*

$$R_C f(x) \leq \varphi(x) \|f\| \tag{5.14}$$

for all $x \in E$ and $f \in \mathfrak{N}^+$.

Then

$$\mathfrak{M}_C \subset \mathfrak{M}_{Cn} = \mathfrak{M}_0, \tag{5.15}$$

where the inclusion is dense and the operators

$$S = R - P_n R_C (I - \Pi), \qquad S' = R - (I - \Pi) R_C P_n \tag{5.16}$$

may be extended to bounded operators acting from $\mathfrak{M}$ to $\mathfrak{M}$.

REMARK 5.2. We see from (5.14) that if the measure μ has a moment of the order φ, that is

$$\int \varphi(x)\, |\mu|(dx) < \infty,$$

then the inclusion $\mu R_C \in \mathfrak{M}$ is valid. If $\mu \in \mathfrak{M}^+$ and (5.14) is, in fact, an equality for a function $f \in \mathfrak{N}^+$ then $\mu R_C \in \mathfrak{M}$ if and only if $\mu\varphi < \infty$.

PROOF. Define kernels

$$I_C(x, A) = \mathbb{I}_{\{x \in A \cap C\}}, \qquad I^C = I - I_C,$$
$$T_C(x, A) = P^n(x, A \cap (E \setminus C)) = P^n I^C(x, A).$$

Let us prove first that the family of measures $(R_C(x, \cdot), x \in E)$ defined in (5.12) be σ-finite uniformly on x. Indeed, the function $\mathsf{E}_x \tau_C = R_C \mathbb{I}(x) \leq \varphi(x)\|\mathbb{I}\|$ is finite by condition (B). Consider a system of sets $A_m = \{x: \mathsf{E}_x \tau_C \leq m\}$. Since $x \in A_m$ implies $\mathsf{P}_x(\tau_C \leq 2m) \geq 1/2$, $\mathbb{I}_{A_m}(x) \leq 2\mathsf{P}_x(\tau_C \leq 2m)$ and, for all $x \in E$,

$$\begin{aligned} R_C(x, A_m) &= \sum_{t \geq 0} \mathsf{P}_x(\tau_C > t, X_{nt} \in A_m) \leq 2 \sum_{t \geq 0} \mathsf{E}_x \left(\mathbb{I}_{\tau_C > t} \mathsf{P}_{X_{nt}}(\tau_C \leq 2m)\right) \\ &= 2 \sum_{t \geq 0} \mathsf{P}_x(t < \tau_C \leq t + 2m) \leq 4m. \end{aligned} \tag{5.17}$$

Here we used the Markov property for τ_C. Estimate (5.17) yields, in particular, that the measure μR_C is σ-finite with respect to the system A_m for all finite measures μ.

Define the subspace

$$\mathfrak{N}_C = \bigcup_{m \geq 1} \{g \in \mathfrak{N}: g(x) = 0, x \notin A_m, \sup_{A_m} |g| < \infty\}. \tag{5.18}$$

Note that $R_C g \in \mathfrak{N}$ for all $g \in \mathfrak{N}_C$, since (5.17) and (5.18) imply that $R_C g$ is bounded and the space $\mathfrak{N}$ contains all bounded functions because of conditions (M1)–(M3), (M$'$).

LEMMA 5.2. *Let μ be a σ-finite measure such that $|\mu g| \leq a\|g\|$ for all $g \in \mathfrak{N}_C$. Then $\mu \in \mathfrak{M}$ and $\|\mu\| \leq a$. In particular, it follows from $\mu g = 0$ for all $g \in \mathfrak{N}_C$ that $\mu = 0$.*

PROOF. We represent μ in the form $\mu = \mu_1 - \mu_2$, where $\mu_i \geq 0$ and $\mu_1 \perp \mu_2$. Choosing $g \in \mathfrak{N}_C$ to be zero outside the supports of μ_1 and μ_2 we make sure that the assertion of the lemma needs to be checked only for $\mu \geq 0$.

Let $g \in \mathfrak{N}^+$. Set $B_k = \{g \leq k\}$. Then $B_k \uparrow E$, since g is finite. Let

$$g_{km}(x) = \begin{cases} g(x), & \text{if } x \in A_m \cap B_k, \\ 0, & \text{otherwise,} \end{cases}$$

where A_m is defined as above. Then $g_{km} \in \mathfrak{N}_C$ and $g_{km} \uparrow g$ as $k, m \to \infty$.

Condition (M2$'$) yields $\|g_{km}\| \leq \|g\|$ and $\mu g_{km} \leq a\|g_{km}\|$ by the condition of the lemma. Therefore, $\mu g_{km} \leq a\|g\|$, where $\mu g_{km} \uparrow \mu g$ by the monotone convergence theorem. Thus, $\mu g \leq a\|g\|$ for all $g \in \mathfrak{N}^+$, whence $|\mu g| \leq a\|g\|$ for all $g \in \mathfrak{N}$, since μ is nonnegative and $\|g_1 - g_2\| \geq \max(\|g_i\|)$ for $g_1 g_2 = 0$ according to conditions (M1)–(M3). So, the measure μ induces a linear continuous functional μg on $\mathfrak{N}$, i.e., μ can be extended to an element of the dual space $\mathfrak{N}^* = \mathfrak{M}^{**}$.

It remains to observe that the measure μ is countable additive because of its σ-finiteness. But the space $\mathfrak{M}^{**}\backslash\mathfrak{M}$ consists of finite measures with a nontrivial purely finite-additive components. Then $\mu \in \mathfrak{M}$ and by Theorem 4.3, [see Rudin (1973)], $\|\mu\| = \sup\left(|\mu g|, \|g\| \leq 1\right) \leq a$. The lemma is proved. □

Lemma 5.2 implies, in particular, the assertion of Remark 5.2. Indeed, if, for $\mu = \mu^+ - \mu^- \in \mathfrak{M}$, $\mu^\pm \varphi < \infty$ then

$$|\mu R_C g| \leq \left(\mu^+ + \mu^-\right) R_C |g| \leq \left(\mu^+ + \mu^-\right) \varphi \|g\| = a\|g\|$$

for $g \in \mathfrak{N}_C$, whence $\mu R_C \in \mathfrak{M}$. Moreover, if $\mu R_C \in \mathfrak{M}$ and $\mu \in \mathfrak{M}^+$ then $(\mu\varphi)\|g\| = \mu R_C g < \infty$ for some $g \in \mathfrak{N}^+$, $\mu\varphi < \infty$.

Let $\mu \in \mathfrak{M}^+$, $f \in \mathfrak{N}^+$. Then by the definition of kernels R_C and T_C we obtain

$$\begin{aligned}\mu R_C f &= \lim_{t\to\infty} \sum_{s\leq t} \mu T_C^s f = \mu f + \lim_{t\to\infty} \sum_{s<t} \mu T_C T_C^s f = \mu f + \mu T_C R_C f \\ &= \mu f + \mu R_C T_C f.\end{aligned}$$

Since the left and right-hand sides of these relations are linear with respect to μ and f,

$$T_C R_C = R_C T_C = R_C - I. \tag{5.19}$$

Consider a stochastic kernel Q concentrated on C

$$Q(x, A) = \mathsf{P}_x\left(X_{n\tau_C} \in A\right). \tag{5.20}$$

Inasmuch as

$$\begin{aligned}\mu R_C P^n I_C f &= \sum_{t\geq 0} \int \mu(dx)\, \mathsf{E}_x\left(P^n I_C f(X_{nt}), \tau_C > t\right) \\ &= \sum_{t\geq 0} \int \mu(dx)\, \mathsf{E}_x\left(f(X_{nt+n}), X_n \notin C, \ldots, X_{nt} \notin C, X_{nt+n} \in C\right) \\ &= \sum_{t\geq 0} \int \mu(dx)\, \mathsf{E}_x\left(f\left(X_{n\tau_C}, \tau_C = t+1\right)\right) = \mu Q f\end{aligned}$$

for $\mu \in \mathfrak{M}^+$ and $f \in \mathfrak{N}^+$ because of the monotone convergence of series, the kernel Q satisfies the operator identity

$$R_C P^n I_C = Q. \tag{5.21}$$

Let a measure μ be concentrated on C. Let us show that $\mu R_C \in \mathfrak{M}$. For $g \in \mathfrak{N}_C$,

$$|\mu R_C g| \leq \int_C |\mu|(dx) |R_C g(x)| \leq \int_C |\mu|(dx)\varphi(x)\|g\| \leq |\mu|(C) M_\varphi \|g\|,$$

where condition (B) is used. Since the measure μR_C is σ-finite, as it is noted above, we conclude from the last estimate and Lemma 5.2 that $\mu R_C \in \mathfrak{M}$ for $|\mu|(E\backslash C) = 0$ and

$$\|\mu R_C\| \leq M_\varphi |\mu|(E). \tag{5.22}$$

Denote by R_n the generalized potential of a chain with the kernel P^n. The domain of R_n is the subspace $\mathfrak{M}_{0n} = \mathfrak{M}\left(I - P^n + \Pi\right)$, where $\Pi = \mathbb{I} \circ \pi$ coincides with the stationary projector Π of the kernel P, since the chain X is aperiodic.

LEMMA 5.3. *Under conditions of Theorem 5.5,*

$$\mathfrak{M}_{0n} = \mathfrak{M}_C \tag{5.23}$$

and the operator

$$S_n = R_n - (I - \Pi) R_C (I - \Pi) \tag{5.24}$$

is bounded on $\mathfrak{M}_{0n}$.

PROOF. Let $\mu \in \mathfrak{M}_{0n}$, that is, $\mu = \nu\,(I - P^n + \Pi)$. Then $\mu(I - \Pi) = \nu\,(I - P^n) = \nu - \nu T_C - \nu P^n I_C$. Thus for all $g \in \mathfrak{N}_C$, so that $R_C g \in \mathfrak{N}$, it follows from (5.19) that

$$\begin{aligned} \mu(I - \Pi) R_C g &= \nu R_C g - \nu T_C R_C g - \nu P^n I_C R_C g \\ &= \nu R_C g - \nu\,(R_C - I)\,g - \nu P^n I_C R_C g = (\nu - \nu P^n I_C R_C)\,g. \end{aligned} \tag{5.25}$$

Since $\nu \in \mathfrak{M}$ and the measure $\nu P^n I_C$ is concentrated on C and is finite, so that $\nu P^n I_C R_C \in \mathfrak{M}$, the factor by g on the right-hand side of (5.23) belongs to $\mathfrak{M}$. Besides, the measure $\mu(I - \Pi) R_C$ on the left-hand side (5.25) is σ-finite. Then from (5.25) and Lemma 5.2 we conclude that $\mu(I - \Pi) R_C \in \mathfrak{M}$, so the left-hand side (5.25) is contained in the right-hand side.

Now we assume that $\mu(I - \Pi) R_C \in \mathfrak{M}$ and show that $\mu \in \mathfrak{M}_{0n}$. Consider a measure

$$\nu = b\pi + \mu(I - \Pi) Q R_q R_C + \mu(I - \Pi) R_C, \tag{5.26}$$

where b is a constant and R_q is the potential of a chain on $(C, C \cap \mathcal{E})$ with the transition kernel $I_C Q$. This chain has a unique invariant measure

$$\pi_C(A) = \pi(A \cap C)/\pi(C),$$

is aperiodic and uniformly ergodic with respect to the total variation norm $\|\mu\| = |\mu|(E)$, since the kernel Q satisfies on C the uniform mixing condition

$$Q(x, A) = Q(x, A \cap C) \geq P^n(x, A \cap C) \geq r\beta(A)$$

according to (5.1) and (5.20). Then the operator R_q transforms finite measures on C into finite measures:

$$|\mu R_q|(C) \leq r_q |\mu|(C) \tag{5.27}$$

for any measure μ on C. One can choose, for example, $r_q = 2/r$ by Theorem 3.4.

Note that the measure $\mu(I - \Pi) Q$ is concentrated on C and is finite. Therefore, the measure $\mu(I - \Pi) Q R_q$ possesses the same property. Then $\mu(I - \Pi) Q R_q R_C \in \mathfrak{M}$ as is shown above while proving (5.22). Hence, the measure ν in (5.26) belongs to $\mathfrak{M}$. Choose the constant b in such a way that $\nu(E) = 0$. Then $\nu\Pi = 0$ and putting $\mu_0 = \mu(I - \Pi)$ and using (5.19), (5.21), we arrive at

$$\begin{aligned} &\nu\,(I - P^n + \Pi) \\ &\quad = \nu\,(I - P^n) = (\mu_0 Q R_q R_C + \mu_0 R_C)\,(I - T_C - P^n I_C) \\ &\quad = (\mu_0 Q R_q + \mu_0) R_C (I - T_C) - (\mu_0 Q R_q R_C P^n I_C + \mu_0 R_C P^n I_C) \\ &\quad = \mu_0 Q R_q + \mu_0 - \mu_0 Q R_q Q - \mu_0 Q = \mu_0 Q R_q (I - Q) + \mu_0 (I - Q) \\ &\quad = \mu_0 Q (I_C - \Pi_C) + \mu_0 (I - Q) = \mu_0 - \mu_0 Q \Pi_C = \mu_0 = \mu(I - \Pi), \end{aligned} \tag{5.28}$$

where the kernel $\Pi_C(x, A) = \pi_C(A)$ is the stationary projector of Q. The identity $R_q(I - Q) = I_C - \Pi_C$ follows immediately from the definition of the potential R_q. According to the definition of Q, Π_C, and Π we have $\mu_0 Q \Pi_C = \mu_0 Q(C)\pi_C = (\mu(E) - \mu\Pi(E))\,\pi_C = 0$.

From (5.28) we get $\mu_0 \in \mathfrak{M}_{0n}$ and $\mu = \mu_0 + (\mu\mathbb{I})\pi \in \mathfrak{M}$, so the equality (5.23) is proved.

Moreover, it follows from (5.28) that the measure ν defined by (5.26) is equal to μR_n. Multiplying (5.26) by $I - \Pi$ and taking into account $\pi(I - \Pi) = 0$ and $\nu\Pi = 0$ we conclude that

$$\mu R_n = \mu(I - \Pi)QR_qR_C(I - \Pi) + \mu(I - \Pi)R_C(I - \Pi)$$

for all $\mu \in \mathfrak{M}_{0n}$. Therefore, the operator S_n given by (5.24) coincides on its domain of definition $\mathfrak{M}_{0n}$ with the operator

$$S_n = (I - \Pi)QR_qR_C(I - \Pi). \tag{5.29}$$

Using consequently estimates (5.22), (5.27), and the inequality

$$|\mu(I - \Pi)Q|(C) \le |\mu(I - \Pi)|(E) \le 2|\mu|(E) \le 4k\|\mu\|$$

followed from (M1′), we obtain

$$\begin{aligned} \|\mu S_n\| &\le \|\mu(I - \Pi)QR_qR_C\|\sigma_1 \le M_\varphi|\mu(I - \Pi)QR_q|(C)\sigma_1 \\ &\le M_\varphi|\mu(I - \Pi)Q|(C)r_q\sigma_1 \le 4kM_\varphi r_q\sigma_1\|\mu\| \end{aligned} \tag{5.30}$$

for $\mu \in \mathfrak{M}_{0n}$, where $\sigma_1 = \|I - \Pi\|$. From (5.30) immediately follows that S_n is bounded on $\mathfrak{M}_{0n}$. The lemma is proved. □

LEMMA 5.4. *Under conditions of Theorem* 5.5,

$$\mathfrak{M}_0 = \{\mu \in \mathfrak{M}: \mu P_n \in \mathfrak{M}_{0n}\} \tag{5.31}$$

and

$$R = P_nR_n, \qquad R = R_nP_n \tag{5.32}$$

on the space $\mathfrak{M}_{0n} \subset \mathfrak{M}_0$ *that is, as well as* $\mathfrak{M}_0$, *dense in* $\mathfrak{M}$.

PROOF. It is easily seen that

$$P_n(I - P) = (I - P)P_n = I - P^n, \qquad P_n\Pi = \Pi P_n = 0.$$

Let $\mu \in \mathfrak{M}_0$, that is, $\mu = \nu(I - P + \Pi)$, $\nu \in \mathfrak{M}$. Then $\mu P_n = \nu(I - P + \Pi)P_n = \nu\,(I - P^n) = \nu(I - \Pi)\,(I - P^n + \Pi) \in \mathfrak{M}_{0n}$. Thus the left-hand side of (5.31) is included in the right-hand side.

Clearly, $\mathfrak{M}_{0n} = \mathfrak{M}\,(I - P^n + \Pi) = \mathfrak{M}(P_n + n\Pi)(I - P + \Pi) \subset \mathfrak{M}_0$ in view of the boundedness of P_n and Π. For $n = 1$, we have $P_1 = I - \Pi$, thus $\mathfrak{M}_{0n} = \mathfrak{M}_0$ and equality (5.31) is satisfied.

If $\mu P_n \in \mathfrak{M}_{0n}$ for $n > 1$ then $\mu P_n = \nu(I - P + \Pi)$ for some $\nu \in \mathfrak{M}$, whence

$$n\mu = \left[\mu\left(n\Pi + \left((n-1)I + \cdots + P^{n-2}\right)(I - \Pi)\right) + \nu\right](I - P + \Pi).$$

Therefore, $\mu \in \mathfrak{M}_0$ and (5.31) is proved.

Equality (5.31) means that domains of operators in the first relation of (5.32) coincide. For this reason, it is sufficient to check that the measure $\mu = \nu R$ coincides with $\nu P_n R_n$ for $\nu \in \mathfrak{M}_0$. By Definition 1.3 of the potential we have $\mu(I - P) = \nu(I - \Pi)$, $\mu\Pi = 0$, whence $\mu(I - P^n) = \mu(I - P)P_n = \nu P_n = \nu P_n(I - \Pi)$. So, $\mu = \nu P_n R_n$ by the definition of R_n and the first equality in (5.32) is proved.

To prove the second one we assume that $\mu = \nu R_n$, $\nu \in \mathfrak{M}_{0n}$. Then $\mu P_n(I-P) = \mu(I - P^n) = \nu(I - \Pi)$, $\mu P_n \Pi = 0$, that is, $\mu P_n = \nu R$ or $\nu R_n P_n = \nu R$ for all $\nu \in \mathfrak{M}_{0n}$, which proves (5.31).

Since $\mathfrak{M}_{0n} \subset \mathfrak{M}_0$, it remains only to prove that the inclusion $\mathfrak{M}_{0n} \subset \mathfrak{M}$ is dense. Assuming the inverse, we find by the Hahn–Banach theorem an element $g \in \mathfrak{N}$ such that $\mathfrak{M}_{0n} g = \{0\}$ and $g \neq 0$. Then $(I - P^n + \Pi)g = 0$. Multiplying the latter equality by Π we obtain $\Pi g = 0$, $g = P^n g$. This relation yields that $(I - T_C)g = P^n I_C g$, whence $g = R_C P^n I_C g = Qg$ because of (5.19) and (5.20).

Since a chain with the kernel Q is aperiodic and uniformly ergodic, we derive that the restriction of g on C is a constant. Since $Q(x, C) = 1$ for $x \in E$, g is constant on the whole E. Since $\Pi g = 0$, $g = 0$. Thus we arrive at a contradiction. The lemma is proved. □

To complete the proof of the theorem, it remains to note that according to (5.23), Lemma 5.4, and definition (5.13)

$$\mathfrak{M}_C = \mathfrak{M}_{0n} \subset \mathfrak{M}_0 = \{\mu \in \mathfrak{M} \colon \mu P_n \in \mathfrak{M}_{0n}\} = \{\mu \in \mathfrak{M} \colon \mu P_n \in \mathfrak{M}_C\} = \mathfrak{M}_{Cn}$$

and the inclusion is dense. Relations (5.15) are proved.

Next, by definition (5.24) of the operator S_n and by Lemma 5.4, operators in (5.16) satisfy the following identities

$$S = P_n S_n, \qquad S' = S_n P_n. \tag{5.33}$$

Here S_n coincides on $\mathfrak{M}_{0n}$ with the bounded operator (5.29), as is shown in (5.30). The operators S and S' are defined on the spaces $\{\mu \in \mathfrak{M} \colon \mu P_n \in \mathfrak{M}_{0n}\} = \mathfrak{M}_0$ and $\mathfrak{M}_{0n}$ respectively. Each of these spaces is dense in $\mathfrak{M}$ by Lemma 5.4. The operator S_n is bounded on $\mathfrak{M}_{0n}$ by Lemma 5.3 and the operator P_n is bounded on whole space. Therefore the operators S and S' admit bounded continuations on $\mathfrak{M}$. The theorem is proved. □

Moreover, as in proof of (5.29), it follows from (5.33) and representation (5.29) that the following explicit estimate is valid

$$\begin{aligned} \|\mu S\| = \|\mu P_n Q R_q R_C (I - \Pi)\| &\le \sigma_1 \|\mu P_n Q R_q R_C\| \le \sigma_1 M_\varphi |\mu P_n Q R_q|(C) \\ &\le \sigma_1 M_\varphi r_q |\mu P_n Q|(C) \le \sigma_1 M_\varphi r_q |\mu P_n|(E) \end{aligned} \tag{5.34}$$

for all $\mu \in \mathfrak{M}_{0n}$, where the obvious estimate $|\nu Q|(C) \le |\nu|(E)$ and inequalities (5.22), (5.27) were used with $\sigma_1 = \|I - \Pi\|$.

Now we show that the condition $\pi D \in \mathfrak{M}_0$ of the Theorem 5.1 is essential for representation (5.3) under assumptions of Theorem 5.5.

THEOREM 5.6. *Let condition* (B) *be satisfied and for some family of stochastic kernels* $Q_\varepsilon = P + \varepsilon B$ *with* $B \in \mathcal{B}$ *fixed, the set* $\mathcal{N}(Q_\varepsilon)$ *is nonempty for any small* $\varepsilon > 0$ *and there exist a family* $\{\nu_\varepsilon\}$ *such that* $\nu_\varepsilon \in \mathcal{N}(Q_\varepsilon)$ *and*

$$\|\nu_\varepsilon - \pi\| = O(\varepsilon) \tag{5.35}$$

as $\varepsilon \to 0$. *Then* $\pi B \in \mathfrak{M}_0$ *and*

$$(\nu_\varepsilon - \pi - \varepsilon\pi BR)g = O(\varepsilon^2) \tag{5.36}$$

on a set of functions g *that is weakly dense in* $\mathfrak{N}$.

PROOF. Let the space $\mathfrak{N}_C$ be defined by (5.18). By Theorem 5.1 for all $g \in \mathfrak{N}_c$, $\varepsilon\nu_\varepsilon BR_g = (\nu_\varepsilon - \pi)g$, where $Rg \in \mathfrak{N}$ by Theorem 5.5, since $R_C g \in \mathfrak{N}$. Then from (5.35) we obtain $\varepsilon|\pi BRg| \leq a\varepsilon\,(\|g\| + \varepsilon\|BR_g\|)$. By Theorem 5.5, the measure $\pi BR = \pi BS' + \pi BR_C P_n$ is σ-finite, since $\pi B \in \mathfrak{M}$. It follows from Lemma 5.1 and from the inequality $|\pi BRg| \leq a\|g\|$ valid for $g \in \mathfrak{N}$ that $\pi BR \in \mathfrak{M}$. Thus $\pi B \in \mathfrak{M}_0$ by Theorem 1.2.

Relation (5.36) for all functions $g \in \mathfrak{N}_C$ immediately follows from (5.3) with $D = \varepsilon B$ and $(\nu_\varepsilon - \pi - \varepsilon\pi BR)g = \varepsilon^2\nu_\varepsilon BRBRg = \varepsilon(\nu_\varepsilon - \pi)BRg = o(\varepsilon^2)$ because $Rg \in \mathfrak{N}$. The theorem is proved. □

5.3. Induced norm of a perturbation

By Theorem 5.5, for any measure $\mu \in \mathfrak{M}_C$ with $\mu(E) = 0$, the following norm is well defined

$$\|\mu\|_C = \|\mu R_C\| \tag{5.37}$$

because of $\mu R_C = \mu(I - \Pi)R_C \in \mathfrak{M}$.

REMARK 5.3. The norm defined by (5.37) is such that

$$\|\mu\|_C \leq |\mu|\varphi \tag{5.38}$$

where $|\mu|$ is the total variation of μ and the function φ is involved into condition (B). Indeed, condition (B) of Lemma 5.2 implies that $\mu R_C g \leq (\mu\varphi)\|g\|$ for $\mu \in \mathfrak{M}^+$, $g \in \mathfrak{N}^+$, whence $\|\mu R_C\| \leq \mu\varphi$. Next, $\|\mu R_C\| \leq \|\,|\mu| R_C\,\|$ because R_C is nonnegative and (M2)–(M3) are satisfied.

If a function $f \in \mathfrak{N}^+$ satisfies (5.14) and $\mu \in \mathfrak{M}^+$ then $\|\mu\|_C = \mu\varphi$, since $\|f\| \cdot \|\mu\|_C \geq \mu R_C f = (\mu\varphi)\|f\|$.

Using relations (5.23) and (5.13) and identity $P_n\Pi = 0$ we get $\mu P_n \in \mathfrak{M}_C$ and $\mu P_n\Pi = 0$ for all $\mu \in \mathfrak{M}_0$. Therefore, the norm $\|\mu\|_0$ on $\mathfrak{M}_0$ is well defined by

$$\|\mu\|_0 = \|\mu P_n\|_C + |\mu(E)| = \|\mu P_n R_C\| + |\mu(E)| \tag{5.39}$$

provided that the kernel P satisfies condition (5.1).

THEOREM 5.7. *Let condition* (B) *hold true. Then the space* $(\mathfrak{M}_0, \|\cdot\|)$ *is Banach and the linear mapping* $R: \mathfrak{M}_0 \to \mathfrak{M}$ *acting from* $(\mathfrak{M}_0, \|\cdot\|_0)$ *into* $(\mathfrak{M}, \|\cdot\|)$ *is bounded.*

REMARK 5.4. The norm of the operator $R: \mathfrak{M}_0 \to \mathfrak{M}$ may be estimated as follows

$$\|R\|_{\mathfrak{M}_0,\mathfrak{M}} \leq \sigma_1(1 + 4kM_\varphi r_q), \tag{5.40}$$

where $\sigma_1 = \|I - \Pi\|$, $M_\varphi = \sup_{x\in C} \varphi(x)$, $r_q \leq 2/r$ is the norm of the potential of the chain $(X_{nt}, t \geq 0)$ observed only at the hitting moments of the set C, the constants n and r are involved in (5.1), and the constant k is taken from condition (M1).

PROOF. Evidently, $\|\mu\|_0$ in (5.39) possesses the property for a norm of a sum. If $\|\mu\|_0 = 0$ then $\mu\mathbb{I} = 0$ and $\mu P_n R_C = 0$, whence $\mu P_n = \mu P_n R_C(I - T_C) = 0$. So $\mu(I - P^n) = \mu P_n(I - P) = 0$ and $\mu = a\pi$ because the chain X is aperiodic. Then $a = \mu\mathbb{I} = 0$, so that $\mu = 0$.

It remains to check that $\mathfrak{M}_0$ is complete with respect to the norm under consideration. Let a sequence $\mu_k \in \mathfrak{M}_0$ be fundamental in the norm $\|\cdot\|_0$. Then $\mu_k\mathbb{I} - \mu_l\mathbb{I} \to 0$ and $\|(\mu_k - \mu_l)P_n R_C\| \to 0$. Setting $a_k = \mu_k\mathbb{I}$, $\alpha_k = \mu_k P_n$, we conclude that there exists the limit $a = \lim_{k\to\infty} a_k$ and the sequence of measures $\nu_k = \alpha_k R_C$ converge in $\mathfrak{M}$ to a measure $\nu \in \mathfrak{M}$. Since $\alpha_k = \alpha_k R_C(I - T_C) = \nu_k(I - T_C)$ because of $\nu_k = \alpha_k R_C$, $\alpha_k \to \alpha = \nu(I - T_C)$ in the norm $\|\cdot\|$ by $T_C \in \mathcal{B}$. Here $\alpha\mathbb{I} = 0$ in view $\alpha_k\mathbb{I} = 0$. Therefore $\nu g = \nu(I - T_C)R_C g = \alpha R_C g$ for all $g \in \mathfrak{N}_C$. Here $\nu \in \mathfrak{M}$ and the measure αR_C is σ-finite. Then Lemma 5.1 implies that $\alpha R_C \in \mathfrak{M}$ and $\alpha R_C = \nu$. Thus $\alpha \in \mathfrak{M}_C = \mathfrak{M}_{0n}$ by definition (5.13) and Lemma 5.3. According to (5.31) the equation $\mu_0 P_n = \alpha$ has a unique solution $\mu_0 \in \mathfrak{M}_0$. Consider a measure $\mu = \mu_0 + a\pi$ and evaluate

$$\begin{aligned}\|\mu_k - \mu\|_0 &= |\mu_k\mathbb{I} - \mu\mathbb{I}| + \|(\mu_k - \mu)P_n R_C\| = |a_k - a| + \|\alpha_k R_C - \alpha R_C\| \\ &= |a_k - a| + \|\nu_k - \nu\|.\end{aligned}$$

Since the right-hand side of this equality tends to zero and $\mu \in \mathfrak{M}_0$, $\mu_k \to \mu_0$ with respect to the norm $\|\cdot\|_0$ and the space $(\mathfrak{M}_0, \|\cdot\|_0)$ is complete.

To prove that R is bounded and that estimate (5.40) is valid we assume that $\mu \in \mathfrak{M}_0$. Then $\nu = \mu P_n R_C \in \mathfrak{M}$ and we arrive at

$$\|\mu R\| \leq \|\mu P_n R_C(I - \Pi)\| + \|\mu S\| \leq \sigma_1\|\mu P_n R_C\| + \sigma_1 M_\varphi r_q |\mu P_n|(E) \tag{5.41}$$

because of (5.14) and (5.34). Note that the measure $\mu P_n = \mu P_n R_C(I - T_C) = \nu - \nu T_C$. Since the kernel T_C is substochastic, $|\mu P_n|(E) \leq 2|\nu|(E) \leq 4k\|\nu\| = 4k\|\mu P_n R_C\|$. Substituting this inequality into (5.41) and using definition (5.39), we get the estimate $\|\mu R\| \leq \|\mu\|_0(1 + 4kM_\varphi r_q)\sigma_1$ that is equivalent to (5.40). The theorem is proved. □

Let D be a linear operator on $\mathfrak{M}$ such that $\mathfrak{M}D \subset \mathfrak{M}_0$ and $D(x, E) = 0$. Define

$$\|D\|_0 = \sup(\|\mu D\|_0, \|\mu\| \leq 1) = \|DP_n R_C\|. \tag{5.42}$$

THEOREM 5.8. *Let a perturbation $D = Q - P$ of a kernel P be such that $\mathfrak{M}D \subset \mathfrak{M}_0$ and condition* (B) *be satisfied. Then the index $\|DR\|$ from the Theorem* 5.3 *is such that*

$$\|DR\| \leq \sigma_1(1 + 4kM_\varphi r_q)\|D\|_0, \tag{5.43}$$

where all constants on the right-hand side are defined in Remark 5.2.

In particular, a unique invariant measure ν of the kernel Q is such that

$$\|\nu - \pi\| = O(\|D\|_0) \quad \text{as } \|D\|_0 \to 0.$$

The proof of the theorem immediately follows from (5.40) and the obvious inequality $\|DR\|_{\mathfrak{M},\mathfrak{M}} \leq \|D\|_{\mathfrak{M},\mathfrak{M}_0}\|R\|_{\mathfrak{M}_0,\mathfrak{M}}$, where the first factor on the right-hand side is equal to $\|D\|_0$ by definition (5.42).

The norm $\|\cdot\|_0$ was defined by (5.39) in terms of the kernel R_C which is constructed by the substochastic kernel $T_C(x, A) = P^n(x, A \cap (E \setminus C))$. Let us show that after changing of the kernel T_C on the set C, the norm $\|\cdot\|_0$ is replaced by its equivalent one, so that in assertions of Theorems 5.7 and 5.8 the kernel R_C may be replaced by R_T.

THEOREM 5.9. *Let condition* (B) *hold true and the substochastic kernel T satisfy inequalities*

$$T_C(x, A) \leq T(x, A) \leq P^n(x, A) - r\mathbb{I}_C(x)\beta(A). \tag{5.44}$$

Define a kernel $R_T(x, A) = \sum_{t\geq 0} T^t(x, A)$ and a subspace

$$\mathfrak{M}_T = \{\mu \in \mathfrak{M}: \mu(I - \Pi)R_T \in \mathfrak{M}\}.$$

Then $\mathfrak{M}_T = \mathfrak{M}_C$, the kernel $R_T - R_C$ admits a bounded continuation to $\mathfrak{M}$, and, for all $\mu \in \mathfrak{M}_T$,

$$\frac{1}{a}\|\mu R_C\| \leq \|\mu R_T\| \leq a\|\mu R_C\|, \tag{5.45}$$

where $a = 1 + 4kM_\varphi r^{-1}$.

PROOF. It follows from (5.44) that $T_C(x, A) = P^n(x, A) = T(x, A)$ for all $x \in E$, $A \in \mathcal{E}$, provided $A \cap C = \varnothing$. Consider the kernel $Q_T = R_C(T - T_C)$. This kernel is nonnegative and σ-finite. Inequalities (5.44) and definition of T_C yield that

$$\begin{aligned} Q_T &\leq R_C\left(P^n - r\mathbb{I}_C \circ \beta - P^n I^C\right) = R_C\left(P^n I_C - r\mathbb{I}_C \circ \beta\right) \\ &= Q - rR_C\mathbb{I}_C \circ \beta = Q - r\mathbb{I}_C \circ \beta, \end{aligned} \tag{5.46}$$

where the kernel Q is defined by (5.20) (see, also (5.21)). We used here the equality $R_C\mathbb{I}_C = \mathbb{I}_C$ following from (5.12). Since $Q_T(x, E \setminus C) = 0$ for $x \in E$ and $Q_T(x, C) \leq Q(x, C) - r\beta(C) = 1 - r$ for $x \in C$ because of (5.46), the kernel $R_Q(x, A) = \sum Q_T^t(x, A)$ is finite, $R_Q(x, E \setminus C) = \mathbb{I}_{\{x\notin C\}}$, and

$$R_Q(x, C) \leq \mathbb{I}_{\{x\in C\}} + \int_C Q_T(x, dy) \sum_{t\geq 1} r^t \leq \frac{1}{r} \tag{5.47}$$

for all $x \in E$.

Let $\mu R_C \in \mathfrak{M}$. Then

$$\mu R_C(I-T) = \mu R_C(I - T_C - T + T_C) = \mu R_C(I - T_C) - \mu R_C(T - T_C) = \mu - \mu Q_T$$

because of (5.19). Note that (5.47) and (5.22) imply that $\mu(R_Q - I)R_C \in \mathfrak{M}$ for all $\mu \in \mathfrak{M}$, since $\mu(R_Q - I) = \mu R_Q Q$ is concentrated on C and is finite. So $\mu R_Q R_C \in \mathfrak{M}$. Substituting μR_Q in the identity above instead of μ we obtain

$$\mu R_Q R_C(I-T) = \mu R_Q(I-Q) = \mu. \tag{5.48}$$

Therefore, $\mu \in \mathfrak{M}(I-T)$ and $\mu R_T \in \mathfrak{M}$. Furthermore, it follows from (5.48) that

$$\mu = \mu R_Q R_C(I-T) = \mu(I + QR_Q)R_C(I-T) = \mu R_C(I-T) + \mu Q R_Q R_C(I-T)$$

for $\mu R_C \in \mathfrak{M}$, where the operator QR_QR_C is bounded. Thus

$$R_T = R_C + QR_QR_C \tag{5.49}$$

on the domain of definition of R_C. Since QR_QR_C is bounded, identity (5.49) implies $\mathfrak{M}_T = \mathfrak{M}_C$ and the second assertion of the theorem.

Estimates (5.45) follow from equalities (5.22), (5.47) and from (M1′) as follows

$$\begin{aligned} \|\mu QR_QR_C\| &\le M_\varphi |\mu QR_Q|(C) \le \frac{M_\varphi}{r}|\mu Q|(C) \le \frac{M_\varphi}{r}|\mu|(E) \\ &\le \frac{M_\varphi}{r}|\mu R_C(I - T_C)|(E) \le 2\frac{M_\varphi}{r}|\mu R_C|(E) \le 4k\frac{M_\varphi}{r}\|\mu R_C\|. \end{aligned} \tag{5.50}$$

Substituting (5.50) and a similar inequality with R_C instead of R_T into estimates for $\|\mu R_T\|$ and $\|\mu R_C\|$ that follow from (5.49), we finally obtain (5.45). The theorem is proved. □

COROLLARY 5.1. *Let a norm on $\mathfrak{M}$ be of the form*

$$\|\mu\|_v = \int v(x)\,|\mu|(dx), \qquad \mathfrak{M} = \mathfrak{M}_v = \{\mu \in m\mathcal{E} : \|\mu\|_v < \infty\},$$

where the function v is such that $\inf v > 0$. Assume that there exists a function $\varphi \in f\mathcal{E}^+$ such that $\sup_{x \in C}\varphi(x) < \infty$ and

$$\lim_{t\to\infty} \mathsf{E}_x\left(\varphi(X_{nt}), \tau_C > t\right) = 0,$$

$$v(x) + \mathsf{E}_x\left(\varphi(X_n), X_n \notin C\right) \le \varphi(x) \tag{5.51}$$

for all $x \in E$. Then the norm $\|\cdot\|_C$ in (5.31) is equal to

$$\begin{aligned} &\|\mu\|_C = \sup\left(|\mu f_1 - \mu f_2|, f_i \in \mathfrak{F}_C\right) \le \|\mu\|_\varphi, \\ &\mathfrak{F}_C = \{f \in f\mathcal{E} : 0 \le f \le v + \mathsf{E}.(f(X_n), X_n \notin C) \le \varphi\}, \end{aligned} \tag{5.52}$$

where $\|\cdot\|_\varphi$ is constructed by φ in the same way as $\|\cdot\|_v$ by v. Moreover, the following relations are valid

$$\{\mu \in \mathfrak{M}: \|\mu P_n\|_\varphi < \infty\} \subset \mathfrak{M}_0 = \{\mu \in \mathfrak{M}: \|\mu P_n\|_C < \infty\} \tag{5.53}$$

and

$$\begin{aligned}\|D\|_0 &= \sup\left(\frac{1}{v(x)}\,|DP_n(f_1 - f_2)(x)|\,, x \in E, f_i \in \mathfrak{F}_C\right)\\ &\le \|D\|_\varphi = \sup_{x\in E}\frac{1}{v(x)}\int \varphi(y)\,|DP_n|(x,dy)\end{aligned} \tag{5.54}$$

for any kernel $D: \mathfrak{M}_0 \to \mathfrak{M}$ with $D(x,E) = 0$, $x \in E$. Finally, one can choose $\varphi(x) = \mathsf{E}_x\tau_C$ if $v \equiv 1$ provided this function is bounded on C.

PROOF. It is easily seen that $\mathfrak{F}_C = \{R_C g, 0 \le g \le v\}$. Indeed, properties of φ give us $\varphi = R_C(I - T_C)\varphi \ge R_C v$ and for any $0 \le g \le v$, $R_C g = g + T_C R_C g \le v + T_C R_C g \le v + T_C R_C v$, so that $R_C g \in \mathfrak{F}_C$.

On the other hand, if $f \in \mathfrak{F}_C$ then it is the minimal solution of the equation $f = g + T_C f$ for some $0 \le g \le v$. The minimality is a consequence of (5.51) and the last condition in (5.52), since $T_C^t f \le T_C^t\varphi \to 0$ as $t \to \infty$ for all $x \in E$. This minimal solution is represented in the form $f = \lim \sum_{s<t} T_C^s g$ and by the definition coincides with $R_C g$. So $f = R_C g$.

It remains to use definitions of norms $\|\cdot\|_v$ and $\|\cdot\|_C$:

$$\begin{aligned}\|\mu R_C\|_v &= \sup(|\mu R_C g|, |g| \le v) = \sup(|\mu f|, f = R_C g, |g| \le v)\\ &= \sup(|\mu f_1 - \mu f_2|, f_i \in \mathfrak{F}_C) = \|\mu\|_C,\end{aligned} \tag{5.55}$$

whence (5.52) immediately follows. The estimate $\|\mu\|_C \le \|\mu\|_\varphi$ is obtained from the inclusion $\mathfrak{F}_C \subset \{f: 0 \le f \le \varphi\}$ and the definition of $\|\cdot\|_\varphi$.

The equality (5.54) is also derived from (5.55). We apply Theorem 5.5 and note that $\mu P_n R_C \in \mathfrak{M}_v$ if and only if $\|\mu P_n R_C\| = \|\mu P_n\|_C < \infty$. The inclusion (5.53) follows from the inequality in (5.52).

Relations (5.54) are obtained from (5.52), (5.42), (5.39) by using expression (1.5) for the operator norm. The corollary is proved. □

COROLLARY 5.2. *Let $E = \{0, 1, \dots\}$ and let the norm of a measure $\mu = (\mu_j)$ on E be defined by $\|\mu\| = \sum |\mu_j|$. Let a chain X on E be ergodic with the invariant distribution $\pi = (\pi_j)$, where $\pi_0 > 0$. Denote by $m_{ij} = \mathsf{E}_i\tau_j$ the mean hitting time of a state j from an another state i. Then conditions* (C) *and* (5.51) *are fulfilled for $C = \{0\}$, $n = 1$, and for the function $\varphi(i) = m_{i0}$. In so doing, the norm in (5.52) is equal to*

$$\|\mu\|_0 = \sum_{j\ge 0}\pi_j\left|\sum_{i\ge 0}\mu_i(m_{ij} - m_{i0})\right| \le \sum |\mu_i| m_{i0} \tag{5.56}$$

and the corresponding induced norm of a perturbation $D = (d_{ij})$ in (5.42) and (5.43) is given by

$$\|D\|_0 = \sup_{k\ge 0}\sum_{j\ge 0}\pi_j\left|\sum_{i\ge 0} d_{ki}(m_{ij} - m_{i0})\right| \le \sup_{i\ge 0}\sum_{j\ge 0}|d_{ij}| m_{j0}. \tag{5.57}$$

In particular, in order that the invariant stochastic measure ν of the perturbed kernel $Q = P + D$ exist, it is sufficient that $\|D\|_0 < \frac{1}{2}$.

EXAMPLE 5.1. Let a chain X on E with the transition matrix $P = (p_{ij})$ be irreducible and positive recurrent, and the stochastic matrix $Q = (q_{ij})$ be such that

$$\frac{p_{ij}}{1+\varepsilon_i} \le q_{ij} \le p_{ij}(1+\varepsilon_i)$$

for all i, j and for some $\varepsilon_i \ge 0$, $\varepsilon_i \to 0$ as $i \to \infty$.

Assume that

$$\limsup \varepsilon_i m_{i0} < \frac{1}{2}. \tag{5.58}$$

Then the chain with the transition matrix Q also is positively recurrent.

PROOF OF COROLLARY 5.2. To check (5.56) it is sufficient to note that, for $C = \{0\}$ and for compactly supported g, the solution of the equation $f = g + PI^C f$ is given by $f(i) = \sum g(j)\, \mathsf{E}_i\, \nu_0(j)$ where $\mathsf{E}_i\, \nu_0(j)$ is the mean hitting number of a state j before the first hitting of the zero state. This value is equal, [see Chung (1960), Corollary 2 of Theorem 11.3], to $\pi_j(m_{ij} - m_{i0} + m_{0j})$. Thus (5.56) is derived from (5.52), and (5.57) from (5.54).

To prove the last assertion of the corollary we use Theorem 5.3 and estimate (5.43) with $\sigma_1 = \|I - \Pi\| = 2$ and $r_q = 0$ because the set $C = \{0\}$ contains only a single point and the potential of a Markov chain on C is equal to zero. Clearly that $\|D\|_0 < 1/2$ implies $\|DR\| < 1$, whence, by Theorem 5.3, the desired assertion follows. □

Conditions posed in Example 5.1 on the matrix Q ensure that the corresponding perturbation $d_{ij} = q_{ij} - p_{ij}$ is such that $|d_{ij}| \le \varepsilon_i p_{ij}$. Using estimate (5.57) we obtain that

$$\|D\|_0 \le \sup_{i\ge 0} \sum_{j\ge 0} \varepsilon_i p_{ij} m_{j0} = \sup_{i\ge 0} \varepsilon_i (m_{i0} - p_{i0} + p_{i0} m_{00}). \tag{5.59}$$

It is clear that the property of the positive recurrence does not depend on any change of a finite number of rows of the matrix Q. For this reason one can assume that $\varepsilon_i = 0$ for all $i < i_0$. Taking i_0 sufficiently large, we derive from (5.58) and (5.59) that $\|D\|_0 < \frac{1}{2}$. According to Corollary 5.2, a chain with the kernel Q has a unique invariant stochastic measure. Thus it is positively recurrent, [see Gikhman and Skorokhod (1973), Chapter 2.5], since irreducibility of Q follows from irreducibility of P.

5.4. Stability of finite-time distributions

We assume in this paragraph that condition (B) holds true, and a norm $\|\cdot\|$ on $\mathfrak{M}$ satisfies condition

(M4) $\|\mu_1 + \mu_2\| = \|\mu_1\| + \|\mu_2\|$ for $\mu_i \in \mathfrak{M}^+$

instead of (M2).

Note that condition (M4) is satisfied for all norms $\|\cdot\|_v$ defined by (1.8), and, in particular, for the total variation norm $|\mu|(E)$.

THEOREM 5.10. *Let a measure μ and a kernel D be such that $\overline{\mu}R_C \in \mathfrak{M}$, $R_C\overline{D} \in \mathcal{B}$ for the measure $\overline{\mu} = |\mu - \mu\Pi|$ and the kernel $\overline{D} = |D - \Pi D|$. Then there exists a constant $a = a(P, C, n)$ such that*

$$\sum_{t\geq 0} \left\|\mu \left(P^t - \Pi\right)\right\| \leq a\left\|\overline{\mu}R_C\right\| \tag{5.60}$$

and, for all $\lambda \in \mathfrak{M}^+$,

$$\sum_{t\geq 0} \left\|\lambda \left(P^t - \Pi\right) D\right\| \leq a\|\lambda\| \cdot \|R_C\overline{D}\|. \tag{5.61}$$

PROOF. Observe that

$$\sum_{t\geq 0} \left\|\mu \left(P^t - \Pi\right)\right\| = \sum_{s<n}\sum_{t\geq 0} \left\|\mu \left(P^{nt} - \Pi\right) P^s\right\|, \tag{5.62}$$

$$\sum_{t\geq 0} \left\|\lambda \left(P^t - \Pi\right) D\right\| \leq \sum_{s<n}\sum_{t\geq 0} \left\|\left(\lambda P^s\right) \left(P^{nt} - \Pi\right) D\right\|, \tag{5.63}$$

where $\|\lambda P^s\| \leq b_n\|\lambda\|$ for $s < n$. Then it is sufficient to check (5.60) and (5.61), where the kernel P^n is used on the left hand side instead of P. Since R_C also is constructed by P^n, we assume in the sequel that the constant n in condition (C) and in definitions (5.11), (5.12) is equal to the unity.

Let us introduce a kernel $T = P - h \circ \alpha$ where the function $h(x) = r\mathbb{I}_{\{x\in C\}}$ and $\alpha = \beta$ in condition (C). Note that $T \geq 0$ and $\alpha h = r > 0$. Furthermore, T satisfies condition (5.44), since T coincides with the right-hand side of (5.44) $n = 1$. Thus the domains of operators R_T and R_C defined above coincide.

Consider a nonnegative sequence

$$g_t = \alpha T^{t-1}h.$$

Lemma 3.1 implies $\alpha R_T = \pi/(\pi h)$, $R_T h = \mathbb{I}$. As in Lemma 3.2 one can derive now that (g_t) is a probability distribution with the finite mean $m = \sum t g_t = \alpha R_T^2 \mathbb{I} = 1/\pi h$ and with the second moment $m_2 = \sum t(t-1)g_t = \pi T R_T \mathbb{I}/\pi h \leq \infty$. Denote by $u_t = \sum_{s\geq 0} (g^{*s})_t$ the corresponding renewal sequence generated by (g_t). Since $g_1 = \alpha h > 0$, the distribution (g_t) is aperiodic and

$$u = \sum_{t\geq 0} |u_{t+1} - u_t| \leq \omega_1(\pi h) < \infty \tag{5.64}$$

by Lemma 3.3. Here ω_1 is defined by (3.32) where $r = \alpha h$ is used instead of αh.

LEMMA 5.5. *The following identities are valid*

$$\begin{aligned} P^t - \Pi &= (I - \Pi)T^t - \sum_{i+j\leq t} (u_i - u_{i-1})(I - \Pi)T^j\mathbb{I} \circ \alpha T^{t-i-j} \\ &= T^t(I - \Pi) - \sum_{i+j\leq t} (u_i - u_{i-1})T^j h \circ \pi T^{t-i-j}(I - \Pi). \end{aligned} \tag{5.65}$$

The proof of the lemma immediately follows from representation (3.36), where $T_z\mathbb{I}$, αT_z, and T_z coincide with the generating functions of sequences $T^t\mathbb{I}$, αT^t, and T^t, respectively, and the function

$$(\alpha T_z\mathbb{I})^{-1} = (1-z)(1-z\alpha T_z h)^{-1} = \frac{1-z}{1-g(z)} = (1-z)\sum u_t z^t$$

is equal to the generating function of the sequence $u_t - u_{t-1}$, where, by the definition, we put $u_{-1} = 0$.

To prove estimate (5.60) we multiply the first representation in (5.65) by μ and, applying Lemma 1.1, condition (M4), and (5.64), obtain

$$\begin{aligned}\sum_{t\geq 0} \|\mu\,(P^t - \Pi)\| &\leq \sum \|\mu(I-\Pi)T^t\| \\ &\quad + \sum_{t\geq 0}\sum_{i+j\leq t} |u_i - u_{i-1}|\,|\mu(I-\Pi)T^j\mathbb{I}|\,\|\alpha T^{t-i-j}\| \\ &\leq \sum_{t\geq 0}\|\overline{\mu}T^t\| + u\sum_{j\geq 0}\overline{\mu}T^j\mathbb{I}\sum_{i\geq 0}\|\alpha T^i\| \\ &= \|\overline{\mu}R_T\| + u\,(\overline{\mu}R_T\mathbb{I})\,\|\alpha R_T\| \leq \|\overline{\mu}R_T\|\,(1+\omega_1\|\Pi\|)\,,\end{aligned} \tag{5.66}$$

since for all measures $\lambda \in \mathfrak{M}^+$

$$\sum_{s<t}\|\lambda T^s\| = \left\|\sum_{s<t}\lambda T^s\right\| \leq \|\lambda R_T\| \tag{5.67}$$

uniformly in t.

Substituting estimate (5.45) into (5.66), we obtain (5.60). Similarly, multiplying from the right the second equality in (5.65) by D and from the left by λ we get

$$\begin{aligned}&\sum_{t\geq 0}\|\lambda\,(P^t-\Pi)\,D\| \\ &\quad\leq \sum_{t\geq 0}\|\lambda T^t(I-\Pi)D\| + \sum_{t\geq 0}\sum_{i+j\leq t}|u_i - u_{i-1}\,(\lambda T^j h)\,\|\pi T^{t-i-j}(I-\Pi)D\| \\ &\quad\leq \sum_{t\geq 0}\|\lambda T^t\overline{D}\| + u(\lambda R_T h)\sum_{t\geq 0}\|\pi T^t D\| \\ &\quad\leq \|\lambda R_T\overline{D}\| + u(\lambda R_T h)\|\pi R_T\overline{D}\| \leq \|\lambda\|(1+\omega_1\|\Pi\|)\|R_T\overline{D}\|.\end{aligned} \tag{5.68}$$

Observe that the operator $R_T - R_C$ is bounded by Theorem 5.9. Since $0 \leq \overline{D} \leq R_C\overline{D}$, $\|R_T\overline{D}\| \leq \|R_C\overline{D}\| + b\|\overline{D}\| \leq \|R_C\overline{D}\|(1+b)$. Now from (5.68) we conclude that (5.61) is valid. The theorem is proved. □

THEOREM 5.11. *Let the initial distribution α of chains X and Y be such that $|\alpha - \pi|R_C \in \mathfrak{M}$ (for instance, this condition is satisfied if the integral $\int \varphi(x)\,|\alpha - \pi|(dx)$ is finite). Assume that $\mathfrak{M}D \subset \mathfrak{M}_0$ and $q = \sup\|Q^t\| < \infty$. Then the representation is valid*

$$\sup_{t\geq 0}\|\mathsf{P}_\alpha(Y_t\in\cdot) - \mathsf{P}(X_t\in\cdot)\| = O(\|D\|_0) \tag{5.69}$$

as $\|D\|_0 \to 0$, where the norm $\|\cdot\|_0$ is defined by (5.38)*. Moreover, the convergence in* (5.69) *is uniform in α with bounded norm $\||\alpha - \pi|R_C\|$ and in all perturbations D with bounded q.*

REMARK 5.5. Under condition $\pi R_C \in \mathfrak{M}$, for the uniform by α convergence in (5.69) it is sufficient that the moment $\alpha\varphi = \int \varphi(x)\,\alpha(dx)$ is finite, see Remark 5.3.

PROOF. By Theorem 5.8, the chain Y has a unique invariant stochastic measure ν for sufficiently small $\|D\|_0$. Note that the expression under the norm sign in (5.69) is equal to

$$\begin{aligned} \alpha Q^t - \alpha P^t &= -\alpha(I - \Pi)\left(P^t - Q^t\right) + \alpha\Pi\left(Q^t - P^t\right) \\ &= -\alpha(I-\Pi)\sum_{s=0}^{t-1} P^s(P-Q)Q^{t-1-s} - \pi\left(Q^t - I\right) \\ &= \sum_{s=0}^{t-1} \alpha\left(P^s - \Pi\right) D Q^{t-1-s} + (\pi - \nu)\left(I - Q^t\right), \end{aligned} \tag{5.70}$$

since $\alpha\Pi = (\alpha\mathbb{I})\pi = \pi$, $\pi = \pi P^t$, $\nu = \nu Q^t$. From (5.70) we obtain

$$\sup_{t\ge 0}\left\|\alpha Q^t - \alpha P^t\right\| \le \sum_{s\ge 0}\|\alpha\left(P^s - \Pi\right)\|\, q\|D\| + \|\nu - \pi\|(1+q). \tag{5.71}$$

Here $\|D\| = o(\|D\|_0)$ because of (5.43). Indeed, condition (P), and Definition 1.3 of the potential imply that

$$\|D\| = \|D(I-\Pi)\| = \|DR(I-P)\| \le \|DR\|(1 + \|P\|) = O(\|DR\|) = O(\|D\|_0).$$

Thus the assertion of the theorem immediately follows from estimate (5.60) in Theorem 5.10 and from Theorem 5.8. □

THEOREM 5.12. *Let $q = \sup\|Q^t\| < \infty$ and a perturbation Q be such that the kernel $\overline{D} = |D - \Pi D|$ satisfy the inclusion $R_C\overline{D} \in \mathcal{B}$. Then the following representation holds true*

$$\sup_{t\ge 0}\|\mathsf{P}_\alpha(Y_y \in \cdot) - \mathsf{P}_\alpha(X_t \in \cdot)\| = O\left(\|D\|_0 + \|R_C\overline{D}\|\right), \tag{5.72}$$

provided its right-hand side tends to zero. Moreover, the convergence in (5.72) *is uniform in α with bounded norm $\|\alpha\|$ and in all perturbations D with bounded q.*

PROOF. It is sufficient to use in (5.70) the estimate (5.61) instead of (5.60):

$$\begin{aligned} \sup_{t\ge 0}\left\|\alpha Q^t - \alpha P^t\right\| &\le \sum_{s\ge 0}\|\alpha\left(P^s - \Pi\right) D\|\, q + \|\nu - \pi\|(1+q) \\ &= O(\|\overline{R}_C D\|) + O(\|D\|_0) \end{aligned}$$

uniformly in α. The theorem is proved. □

CHAPTER VI

UNIFORM CONSOLIDATION THEOREMS

The estimates for the closeness of distributions of chains X and Y obtained in the previous Chapter are essentially based on the assumption of the uniqueness of the invariant measure of the basic chain X. Actually, it was our main assumption regarding the chain X. But in the case where the chain X is reducible and so it has several essential classes of states, the distribution of the perturbed chain Y does not approach the distribution of X even if the perturbation $D = Q - P$ is small. In the consolidation theory of Markov process, [Korolyuk and Turbin (1976), (1978)], a form of the chain that approximates the perturbed chain Y is found. This chain is called consolidated.

In this Chapter we obtain estimates for the convergence that are expressed, like those in (5.6), in terms of the potential of the chain X. Note, that the restriction of this potential on each minimal class of states of X also is a potential and thus one can apply to it all results of Chapter 5.

All results of this Chapter are valid for norms on $\mathfrak{M}$ satisfying the condition (M1). This condition is assumed throughout in this Chapter.

6.1. Perturbed, basic, and consolidated chains

Consider a homogeneous Markov chain $Y = (Y_t)$ on $(E, \mathcal{E})$ with a transition kernel $Q(x, A)$. This chain is called initial or perturbed. Assume that its kernel is close to the transition kernel $P(x, A)$ of the basic Markov chain $X = (X_t)$ on $(E, \mathcal{E})$ so that the perturbation $D(x, A) = Q(x, A) - P(x, A)$ is small.

How to choose a basic chain for a given initial one? This is an important special problem that essentially depends on the structure of the initial chain.

Let $(U, \mathcal{U})$ be a measurable space treated in the sequel as a domain of the consolidated chain. Let also $\varphi: E \to U$ be a measurable function taking values in the space $(U, \mathcal{U})$ and $\mathcal{E}_0 \subset \mathcal{E}$ be a sub-σ-field of $\mathcal{E}$. Suppose that φ is $\mathcal{E}_0$-measurable and denote by $\mathfrak{N}\mathcal{E}_0$ the class of $\mathcal{E}_0$-measurable functions on the space $\mathfrak{N}$.

DEFINITION 6.1. A stochastic kernel $\Pi(x, A)$ on $(E, \mathcal{E})$ is $\mathcal{E}_0$-adapted if the function $\Pi(\cdot, A)$ is $\mathcal{E}_0$-measurable for all $A \in \mathcal{E}$ and $f = 0$ provided $\Pi f = 0$, $f \in \mathfrak{N}\mathcal{E}_0$.

It is easily seen that the Markov chain X_t^0 that corresponds to the kernel Π is such that $\varphi\left(X_t^0\right)$ also is a Markov chain on $(U, \mathcal{U})$.

Assume that the basic chain X satisfies the following condition

(O) the kernel $P \in \mathcal{B}$ has a $\mathcal{E}_0$-adapted stationary projector $\Pi \in \mathcal{B}$.

Here we used Definitions 6.1 and 1.2. Under assumptions of Remark 1.3, where the projector Π is defined, condition (O) is satisfied if, for example, $\mathcal{E}_0 = \sigma(E_u, u \in U)$.

As in section 1.2, we define the subspace $\mathfrak{M}_0 = \mathfrak{M}(I - P + \Pi)$ and the generalized potential R corresponding to the basic chain X.

Define on the space $(E, \mathcal{E})$ a consolidated (in other words, approximating) Markov chain $Y^0 = \left(Y_t^0, t \geq 0\right)$ with the transition kernel

$$Q_0(x, A) = \Pi Q \Pi(x, A) = \Pi(x, A) + \Pi D \Pi(x, A) \tag{6.1}$$

that is obtained by means of averaging of the kernel Q. The assertion given below shows that the structure of the consolidated chain Y_t^0 is essentially simpler than that of the perturbed chain Y.

REMARK 6.1. Let the map $u \to \varphi^{-1}(u)$ have a measurable selector $\psi: U \to E$. Then the stochastic sequence $Z_t = \varphi\left(Y_t^0\right)$ with values in $(U,\mathcal{U})$ is a Markov chain that is isomorphic to Y^0. Moreover, for all $t \geq 0$ and $u \in U$, $B \in \mathcal{U}$, $x \in E$, $A \in \mathcal{E}$, the following identities are valid

$$\mathsf{P}\left(Z_t \in B/Z_0 = u\right) = \mathsf{P}\left(Y_t^0 \in \varphi^{-1}(B)/Y_0^0 = \psi(u)\right), \tag{6.2}$$

$$\mathsf{P}\left(Y_t^0 \in A/Y_0^0 = x\right) = \int \mathsf{P}\left(Z_t \in du/Z_0 = \varphi(x)\right)\Pi(\psi(u), A). \tag{6.3}$$

Indeed, the function $Q_0(\cdot, A)$ is $\mathcal{E}_0$-measurable for $A \in \mathcal{E}$, whence $Q_0(\psi(u), \varphi^{-1}(B))$ is a transition kernel on $(U,\mathcal{U})$.

We introduce also the condition

(I) the consolidated chain Y^0 has a unique invariant stochastic measure $\nu_0 \in \mathfrak{M}$.

Provided this condition, we denote by $\Pi_0 = \mathbb{I} \circ \nu_0$ the stationary projector of the kernel Q_0. Consider the space $\mathfrak{M}_{00} = \mathfrak{M}(I - Q_0 + \Pi_0)$ that is the domain of definition of the generalized potential R_0 of the consolidated chain Y^0. The value $\mu = \nu R_0$ of the potential coincides with a unique solution of the system $\mu(I - Q_0) = \nu(\Pi - \Pi_0)$, $\mu\Pi_0 = 0$ unlike (6.8). By the definition of Q_0, R_0, and Π_0 we obtain

$$\begin{gathered}
Q_0\Pi = \Pi Q_0 = Q_0, \qquad \Pi_0 Q_0 = Q_0\Pi_0 = \Pi_0, \\
\Pi_0\Pi = \Pi\Pi_0 = \Pi_0, \qquad R_0\Pi = \Pi R_0 = R_0, \\
R_0(I - Q_0) = (I - Q_0)R_0 = \Pi - \Pi_0, \qquad R_0\Pi_0 = \Pi_0 R_0 = 0.
\end{gathered} \tag{6.4}$$

LEMMA 6.1. *Let the perturbation D satisfy conditions*

(D) $\mathfrak{M}D \subset \mathfrak{M}_0$, $\mathfrak{M}D\Pi = \mathfrak{M}\Pi D\Pi$.

Then the operators DR and DR_0 are well defined on $\mathfrak{M}$ and are bounded.

REMARK 6.2. In order that the second condition in (D) be satisfied, it is sufficient, in particular, that the consolidated chain Y^0 be uniformly ergodic in the norm $\|\cdot\|$.

PROOF. By Lemma 5.1, it is sufficient to prove that $\mathfrak{M}D \subset \mathfrak{M}_{00}$. According to (D), for all $\nu \in \mathfrak{M}$ there exists $\alpha \in \mathfrak{M}$ such that $\alpha\Pi D\Pi = \nu D\Pi$. Then the measure $\mu = \nu D(I - \Pi) - \alpha(\Pi - \Pi_0)$ belongs to $\mathfrak{M}$, since $\nu D \in \mathfrak{M}_0 \subset \mathfrak{M}$. It follows from (6.1) and (6.4) that $\mu(I - Q_0 + \Pi_0) = \nu D$. Thus $\nu D \in \mathfrak{M}_{00}$ for all $\nu \in \mathfrak{M}$. If the chain Y^0 is uniformly ergodic then $\mathfrak{M}_{00} = \mathfrak{M}$ by Theorem 1.3 and DR_0 is well defined for all bounded D.

In what follows we assume that conditions (O), (I), and (D) are satisfied.

6.2. Approximation of the invariant distribution

For all perturbations D satisfying (D) the following indices are well defined by Lemma 6.1

$$\varepsilon = \|DR\|, \qquad r_0 = \|DR_0\|, \qquad \varepsilon_{01} = \|\nu_0 DR\|. \tag{6.5}$$

THEOREM 6.1. *Let conditions* (O), (I), *and* (D) *be fulfilled and* $\varepsilon(1+r_0) < 1$. *Then the perturbed chain* Y *has a unique invariant stochastic measure* $\nu \in \mathfrak{M}$ *that is represented in the form*

$$\nu = \nu_0(I - DR - DRDR_0)^{-1}. \tag{6.6}$$

PROOF. Let $\nu = \nu Q \in \mathfrak{M}$ and $\nu\mathbb{I} = 1$. By the definition of Π_0, $\nu\Pi\Pi_0 = \nu\Pi_0 = (\nu\mathbb{I})\nu_0 = \nu_0$ and $D\Pi_0 = (D\mathbb{I})\nu_0 = 0$. From the definition of R_0 and from the identity $\nu\Pi(I - Q_0) = -\nu\Pi D\Pi = -\nu\Pi D(\Pi - \Pi_0)$ we conclude that $\nu\Pi(\Pi - \Pi_0) = -\nu\Pi DR_0$ and

$$\nu\Pi(I + DR_0) = \nu\Pi - \nu\Pi(\Pi - \Pi_0) = \nu_0. \tag{6.7}$$

Analogously, since ν is invariant and $\nu(I - P) = \nu Q - \nu P = \nu D$, $\nu D \in \mathfrak{M}_0$, $\nu D\Pi = 0$, and

$$\nu(I + DR) = \nu\Pi. \tag{6.8}$$

Therefore (6.7) and (6.8) imply that

$$\begin{aligned} \nu(I - DR - DRDR_0) &= \nu - \nu(I - \Pi) - \nu(I - \Pi)DR_0 \\ &= \nu\Pi(I + DR_0) - \nu D\Pi R_0 = \nu_0. \end{aligned} \tag{6.9}$$

It remains to note that the operator $I - DR - DRDR_0$ has a bounded inverse for $\varepsilon(1+r_0) < 1$, since its inverse coincides with a converged Neuman series. Therefore, (6.6) follows from (6.9).

On the other hand, the measure in (6.6) evidently satisfies (6.9). Multiplying (6.9) by $I-\Pi$, we get $\nu(I-\Pi-DR) = 0$, so $\nu DR = \nu(I-\Pi)$, whence $\nu D(I-\Pi) = \nu(I - \Pi)(I - P) = \nu - \nu P$ and $\nu - \nu Q = -\nu D\Pi$. Substituting into (6.9) the corresponding equality for νDR, we conclude that $\nu\Pi - \nu(I-\Pi)DR_0 = \nu_0$, whence by the definition of R_0 we get

$$\nu\Pi - \nu\Pi D\Pi + \nu D\Pi + \nu\Pi D\Pi = \nu D\Pi = 0.$$

Thus $\nu D\Pi = 0$ and $\nu = \nu Q$, as is shown above. So the measure ν in (6.6) is invariant for Q. The equality $\nu\mathbb{I} = 1$ follows from (6.6). The theorem is proved. □

From Theorem 6.1 one can easily obtain estimates for the approximation of the distribution of ν in terms of the distribution of ν_0 that are similar to those obtained in Theorem 5.3.

THEOREM 6.2. *Under conditions of Theorem* 6.1, *a unique invariant stochastic measure* ν *of the perturbed chain* Y *satisfies for all* $f \in \mathfrak{N}$ *the inequality*

$$\left|\mathsf{E}_\nu f(Y_1) - \mathsf{E}_{\nu_0} f\left(Y_1^0\right)\right| \le \frac{\varepsilon_{01}}{1 - \varepsilon(1+r_0)}(1 + r_0)\|f\|. \tag{6.10}$$

REMARK 6.3. If the kernel P is constant in the scheme of the series then the potential R_0 of the kernel $Q_0 = \Pi + \Pi D\Pi$ may change with the change of the perturbation D. But for perturbations of the form $D = \delta B$, where B is fixed and satisfies the condition (D), the invariant distribution ν_0 and the product DR_0 do not depend on δ. That is why the right-hand side of (6.10) is $O(\delta)$ as $\delta \to 0$ in the case under consideration. Thus (6.10) has a maximal order of approximation.

REMARK 6.4. Equations (6.7) and (6.8) may be used for the consecutive evaluation of the distribution ν by ν_0 with an auxiliary calculation of $\nu\Pi$.

PROOF. Evidently, the difference on the left-hand side of (6.10) is equal to

$$\begin{aligned} \nu f - \nu_0 f &= \nu_0(I - DRDR_0 - DR - I)(I - DR - DRDR_0)^{-1} f \\ &= -\nu_0 DR(I + DR_0)(I - DR - DRDR_0)^{-1} f. \end{aligned} \tag{6.11}$$

Estimating the norms of operators involved in (6.11) and using (6.5) we obtain (6.10). The theorem is proved. □

Recall that the equations in the definition of ν_0 and values $\mu = \alpha DR_0$ are of the form $\nu_0 = \nu_0\Pi Q\Pi$, $\nu_0\mathbb{1} = 1$, and $\mu(I - \Pi Q\Pi) = \alpha D(\Pi - \Pi_0)$, $\mu\Pi_0 = 0$. So, by the definition of Π and by the obvious equalities $\nu_0\Pi = \nu_0$, $\mu = \mu\Pi$ these equations are, respectively, equivalent to

$$\nu_0\Pi D\Pi = 0, \qquad \nu_0 = \nu_0\Pi, \qquad \nu_0\mathbb{1} = 1, \tag{6.12}$$

$$\mu\Pi D\Pi = -\alpha D\Pi, \qquad \mu\mathbb{1} = 0, \qquad \mu = \mu\Pi. \tag{6.13}$$

It is clear that, in the case $D = \delta B$, solutions of (6.12) and (6.13) do not depend on δ, and neither do conditions (D) depend on δ.

6.3. Approximation of finite-time distributions

To estimate the approximation accuracy of finite time distributions of chains Y and Y^0, we introduce additional indices

$$r(\alpha) = \sum_{t\geq 0} \left\|\alpha\left(Q_0^t - \Pi_0\right) DR\right\| + \|\alpha R\|, \qquad \varepsilon_0 = \|D\Pi\|. \tag{6.14}$$

In order that $r(\alpha)$ be finite, it is sufficient, in particular, that $\alpha \in \mathfrak{M}_0$, $\mathfrak{M}D \subset \mathfrak{M}_0$, and the chain Y^0 be uniformly ergodic in the norm $\|\cdot\|$. In general, $r(\alpha)$ may be estimated according to Theorem 5.10.

Furthermore, in the sequel, we assume that the perturbed chain Y satisfies the condition

(K) $q = \sup_{t\geq 0} \|Q^t\| < \infty$.

Condition (K) is satisfied, in particular, for the total variation norm.

THEOREM 6.3. *Let conditions* (O), (I), (D), *and* (K) *be satisfied. Assume that the initial distribution α of the chains Y and Y^0 is such that $r(\alpha) < \infty$ and $\varepsilon(1 + r_0) < 1$. Then for all $f \in \mathfrak{N}\mathcal{E}_0$, the following inequality is valid*

$$\sup_{t\geq 0} \left|\mathsf{E}_\alpha f(Y_t) - \mathsf{E}_\alpha f\left(Y_t^0\right)\right| \leq \frac{\varepsilon_0 r(\alpha) + \varepsilon_{01}(1 + r_0)}{1 - \varepsilon(1 + r_0)} q\|f\|, \tag{6.15}$$

where all terms are defined by (6.5) *and* (6.14). *Note that* $r(\alpha) = 0$ *for* $\alpha = \nu_0$.

The proof of the theorem will be obtained simultaneously with a dual to (6.15) estimate that is applicable also for $\mathcal{E}$-measurable functions f.

To give this estimate we consider the space

$$\mathfrak{N}_0 = (I - P + \Pi)\mathfrak{N}_0$$

where the generalized potential R of the basic chain is defined, like that in Definition 1.3, by the following equalities

$$R(I - P) = I - \Pi, \qquad R\Pi = 0.$$

The potential R is well defined provided additionally that the projector Π satisfies the condition $\Pi f = f$ for all $f \in \mathfrak{N}$ such that $Pf = f$. Then from $(I - P)f_1 = (I - P)f_2$ it follows that $f_2 - f_1 = \Pi(f_2 - f_1)$, so $R(I - P)(f_2 - f_1) = 0$. The condition introduced above means that $I - P + \Pi$ has no annihilators in $\mathfrak{N}$ and is equivalent to condition that the subspace $\mathfrak{M}_0 = \mathfrak{M}(I - P + \Pi)$ is dense in the space $\mathfrak{M}$, [see Theorem 4.12 in Rudin (1973)].

Observe that the product RD is well defined and bounded on $\mathfrak{N}$ and therefore it is defined and bounded on $\mathfrak{M}$ if we assume that $D\mathfrak{N} \subset \mathfrak{N}_0$. Therefore, for all $f \in \mathfrak{N}_0$ and all kernels mentioned above, the following indices are well defined

$$\varepsilon' = \|RD\|, \qquad \varepsilon_0' = \|\Pi D\|,$$

$$r'(f) = \sum_{t \geq 0} \|RD\,(Q^t - \Pi_0)\,f\| + \|Rf\|. \tag{6.16}$$

THEOREM 6.4. *Let* $\mathfrak{M}_0$ *be dense in* $\mathfrak{M}$, $D\mathfrak{N} \subset \mathfrak{N}_0$, $\varepsilon' < 1$, $f \in \mathfrak{N}$, *and* $r'(f) < \infty$. *Then for all* $t \geq 0$ *and* $\alpha \in \mathfrak{M}$

$$\left|\mathsf{E}_\alpha f(Y_t) - \mathsf{E}_\alpha f\left(Y_t^0\right)\right| \leq \varepsilon_t(\alpha, f) + 2\frac{\varepsilon_0' r'(f)}{1 - \varepsilon'} q\|\alpha\Pi\|, \tag{6.17}$$

where

$$\varepsilon_t(\alpha, f) = \sum_{s<t} \|\alpha\,(P^s - \Pi)\,D\|\,q\|f\| + \left|\alpha\,(P^t - \Pi)\,f\right| \tag{6.18}$$

vanishes for $\alpha = \alpha\Pi$.

Note that the left-hand side of (6.17) does not converge to zero for $\alpha \neq \alpha\Pi$ and $f \neq \Pi f$ even if perturbed and basic chains are finite, since kernels Q^t and Q_0^t of chains Y and Y^0 do not become close for small t as $D \to 0$. They converge to P^t and Π, respectively. Therefore (6.18) contains a summand that depends on t and tends to zero only for large t. Methods for the estimation of ergodicity indices in (6.18) are given in Chapters 3 and 5.

REMARK 6.5. In the statements of Theorems 6.3 and 6.4 and in the definitions of the corresponding variables (6.14), (6.16), and (6.18), the domain of the time variable t may change for $0 \leq t \leq t_0$ if t_0 is arbitrary but fixed.

The proofs of Theorems 6.3 and 6.5 are based on explicit representations of differences on the left-hand sides of (6.15) and (6.17). To obtain them we consider the following operators

$$H = \Pi DR(I - DR)^{-1}, \qquad G = R(I - DR)^{-1}, \tag{6.19}$$

and

$$H' = (I - RD)^{-1}RD\Pi, \qquad G' = (I - RD)^{-1}R. \tag{6.20}$$

LEMMA 6.2. *Let either* $\mathfrak{M}D \subset \mathfrak{M}_0$ *and* $\varepsilon < 1$ *or* $D\mathfrak{N} \subset \mathfrak{N}_0$ *and* $\varepsilon' < 1$. *Then*

$$H\Pi = 0, \qquad H(I-Q)(I-\Pi) = \Pi(Q-Q_0), \qquad G(I-Q)(I-\Pi) = I-\Pi, \tag{6.21}$$

and

$$\Pi H' = 0, \qquad (I-\Pi)(I-Q)H' = (Q-Q_0)\Pi, \qquad (I-\Pi)(I-Q)G' = I-\Pi \tag{6.22}$$

respectively.

PROOF. Under conditions of the lemma, we get $(I - DR)^{-1} \in \mathcal{B}$, since DR is bounded, as was shown above, and its norm is less than the unity. So, the operator H is bounded and G is defined on $\mathfrak{M}_0$. The equality $H\Pi = 0$ follows from $R\Pi = 0$. Since

$$\Pi D(I-\Pi) = \Pi(Q-P)(I-\Pi) = \Pi Q(I-\Pi) = \Pi Q - Q_0 = \Pi(Q-Q_0),$$

$H = \Pi DG$ and the second equality in (6.21) follows from the third one. For all $\mu \in \mathfrak{M}_0$, we have $\mu G = \mu R(I - DR)^{-1}$ where $\mu R \in \mathfrak{M}$. Hence $\mu G(I - DR) = \mu R$ and, by the definition of the potential,

$$\begin{aligned}\mu(I-\Pi) = \mu R(I-P) = \mu G(I-P-D(I-\Pi)) = \mu G(I-Q+D\Pi) \\ = \mu G(I-Q-(I-Q)\Pi) = \mu G(I-Q)(I-\Pi).\end{aligned}$$

Therefore, the operator $G(I - Q)(I - \Pi)$ has a bounded continuation on $\mathfrak{M}$ and coincides with $I - \Pi$, which proves (6.21).

The equalities (6.22) may be proved in the same way. □

Let us introduce $\mathcal{B}$-valued sequences $[Q]$, $[Q_0]$, $[P]$, $[Q_0]^0$, $[\Pi]$, $[I]$ with t-th elements given by

$$\begin{aligned}&[Q]_t = Q^t, \qquad [Q_0]_t = Q_0^t, \qquad [P]_t = P^t, \\ &[Q_0]_t^0 = Q_0^t - \Pi_0, \qquad [I]_t = I, \qquad [\Pi]_t = \Pi,\end{aligned} \tag{6.23}$$

respectively, where $Q^0 = P^0 = I$ and $Q_0^0 = \Pi$. We also introduce numerical sequences $\sigma = (\delta_{1t}, t \geq 0)$ and $\delta = (\delta_{0t}, t \geq 0)$.

For two sequences $x = (x_t)$ and $y = (y_t)$ that are either numerical or $\mathfrak{M}$ or $\mathfrak{N}$ or $\mathcal{B}$-valued, we define their convolution by

$$(x * y)_t = \sum_{s=0}^{t} x_s y_{t-s}$$

provided that the elements of x may be multiplied by the elements of y from the right side. We treat in the component-wise way the product of a number, of a measure, of a function or of an operator to an appropriate sequence. In the sequel, we use identities (6.4), definition (6.1) and the evident equalities

$$\Pi D = \Pi(Q-I), \qquad D\Pi = (Q-I)\Pi. \tag{6.24}$$

LEMMA 6.3. *The following identities are valid*

$$(\Pi - \Pi_0)([Q] - [Q_0])\Pi = \sigma * [Q_0]^0 H * (D\Pi[Q] - [Q]D\Pi)\Pi, \tag{6.25}$$

$$\Pi([Q] - [Q_0])(\Pi - \Pi_0) = \Pi(\Pi D[Q] - [Q]\Pi D) * H'[Q_0]^0 * \sigma, \tag{6.26}$$

where the operators H and H' are defined by (6.19) *and* (6.20) *respectively.*

PROOF. By using (6.24) we get

$$\begin{aligned}(D\Pi[Q] - [Q]D\Pi)_t\Pi &= \left(D\Pi Q^t - Q^t D\Pi\right)\Pi = -(I-Q)\Pi Q^t\Pi + Q^t(I-Q)\Pi \\ &= -(I-Q)\Pi Q^t\Pi + (I-Q)Q^t\Pi = (I-Q)(I-\Pi)Q^t\Pi.\end{aligned} \tag{6.27}$$

Initial members of the sequences on the left and right-hand sides of (6.25) are evidently equal to zero, since $\sigma_0 = 0$ and $(\Pi - \Pi_0)(I - \Pi)\Pi = 0$. Using (6.21) and (6.27) we evaluate a member on the right-hand side of (6.25) with index $t \geq 1$ as follows

$$\begin{aligned}&\sum_{s=0}^{t-1}(Q_0^s - \Pi_0)\,H(I-Q)(I-\Pi)Q^{t-1-s}\Pi \\ &\quad= \sum_{s=0}^{t-1}(Q_0^s - \Pi_0)\,\Pi D(I-\Pi)Q^{t-1-s}\Pi \\ &\quad= \sum_{s=0}^{t-1}(Q_0^s - \Pi_0)\,\Pi(Q-Q_0)Q^{t-1-s}\Pi \\ &\quad= (\Pi - \Pi_0)\sum_{s=0}^{t-1}Q_0^s(Q-Q_0)Q^{t-1-s}\Pi \\ &\quad= -(\Pi-\Pi_0)\left(Q_0^t - Q^t\right)\Pi = (\Pi - \Pi_0)([Q]-[Q_0])_t\Pi.\end{aligned}$$

Thus (6.25) is proved. The dual identity (6.26) is derived from (6.22) in the same way. □

LEMMA 6.4. *The following identities are satisfied*

$$(I - \Pi)[Q]\Pi = G(D\Pi[Q] - [Q]D\Pi)\Pi, \tag{6.28}$$

$$\Pi[Q](I - \Pi) = \Pi(\Pi D[Q] - [Q]\Pi D)G', \tag{6.29}$$

$$([Q] - [P])(I - \Pi) = \sigma * [Q] * D([P] - [\Pi]), \tag{6.30}$$

$$(I - \Pi)([Q] - [P]) = ([P] - [\Pi])D * [Q] * \sigma. \tag{6.31}$$

PROOF. Using (6.21) and (6.27), we evaluate

$$G(D\Pi[Q] - [Q]D\Pi)_t\Pi = G(I-Q)(I-\Pi)Q^t\Pi = (I-\Pi)Q^t\Pi,$$

which proves (6.28). Removing the parentheses in the sum that defines the convolution and then substituting $D = Q - P$ and $[P] - [\Pi] = [P](I - \Pi)$ we obtain identity (6.30). The equalities (6.29) and (6.31) are derived in the same way. □

Combining assertions of lemmas above we arrive at the following result.

THEOREM 6.5. *Under conditions of Theorem* 6.3, *the numerical sequence formed by differences from the left-hand side of* (6.15) *is equal to*

$$\alpha([Q]-[Q_0])f = (\nu_0-\nu)([I]-[Q])f + \left(\sigma * \alpha[Q_0]^0 H + (\alpha G)\delta\right) * (D\Pi[Q]-[Q]D\Pi)f, \tag{6.32}$$

where ν *is an invariant measure of the kernel* Q *constructed in Theorem* 6.1.

Moreover, under the conditions of Theorem 6.4, *the sequence formed by differences mentioned above is equal to, after subtraction of the sequence* $\alpha(P^t-\Pi)f$,

$$\alpha([Q]-[Q_0])f - \alpha([P]-[\Pi])f = \alpha([P]-[\Pi])D * [Q]f * \sigma + \alpha\Pi(\Pi D[Q]-[Q]\Pi D) * \left(\sigma * H'[Q_0]^0 f + (G'f)\,\delta\right). \tag{6.33}$$

PROOF. It is clear that $f=\Pi f$ by the condition $f\in\mathfrak{N}\mathcal{E}_0$ of Theorem 6.3. Thus it is sufficient to apply the decomposition

$$\begin{aligned}\alpha([Q]-[Q_0])\Pi f &= \alpha\Pi_0([Q]-[Q_0])\Pi f + \alpha(\Pi-\Pi_0)([Q]-[Q_0])\Pi f + \alpha(I-\Pi)([Q]-[Q_0])\Pi f\\ &= (\nu-\nu_0)([I]-[Q])f + \alpha(\Pi-\Pi_0)([Q]-[Q_0])\Pi f + \alpha(I-\Pi)[Q]\Pi f.\end{aligned}$$

Here we use the equalities

$$\alpha\Pi_0 = (\alpha\mathbb{1})\nu_0 = \nu_0, \qquad \nu_0[Q_0]=\nu_0[I], \qquad \nu[I]=\nu[Q]$$

as well as the relation $(I-\Pi)[Q_0]=0$ that follows from (6.4). It remains to substitute (6.25) and (6.28) into the second and the third summands on the right-hand side of the last identity. Finally we get (6.32).

Decomposition (6.33) follows from the representation

$$\begin{aligned}\alpha([Q]-[Q_0])f - \alpha([P]-[\Pi])f &= \alpha(I-\Pi)([Q]-[P])f + \alpha\Pi([Q]-[Q_0])(I-\Pi)f\\ &\quad + \alpha\Pi([Q]-[Q_0])(\Pi-\Pi_0)f + \alpha\Pi([Q]-[Q_0])\Pi_0 f,\end{aligned} \tag{6.34}$$

where $(I-\Pi)[P]=[P]-[\Pi]$. Note that $[Q_0](I-\Pi)=0$ and $([Q]-[Q_0])\Pi_0 = ([Q]\mathbb{1}-[Q_0]\mathbb{1})\nu_0 = 0$, since kernels Q and Q_0 are stochastic. Substituting expressions (6.31), (6.29) and (6.26) into the right-hand side of (6.34) instead of its first three summands, respectively, we obtain (6.33). The theorem is proved. □

PROOF OF THEOREMS 6.3 AND 6.4. On the space of numerical sequences $x=(x_t)$, we define the norms

$$|x|_0 = \sup_{t\ge 0}|x_t|, \qquad |x|_1 = \sum_{t\ge 0}|x_t|.$$

It is easily seen that for any sequences x and y

$$\begin{aligned}&|x*y|_0 \le |x|_0|y|_1, \qquad |x*y|_1 \le |x|_1|y|_1,\\ &|x_0\bar{y}+x*y|_0 \le |x|_0|y|_1, \qquad \bar{y}_t=\sum_{s>t} y_s.\end{aligned} \tag{6.35}$$

Now let $z = (z_t)$ be a sequence of elements from either $\mathfrak{M}$ or $\mathfrak{N}$ or $\mathcal{B}$. Define the norm

$$|z|_i = |\,\|z\|\,|_i, \qquad \|z\| = (\|z_t\|, t \geq 0)$$

as the value of the norm $|\cdot|_i$ of the numerical sequence of norms $\|z\|$. Using an evident estimate for a sum and a product of elements from $\mathfrak{M}$, $\mathfrak{N}$, $\mathcal{B}$ we make sure that inequalities (6.35) are valid for $\mathfrak{M}$-, $\mathfrak{N}$-, and $\mathcal{B}$-valued sequences. Furthermore, for all elements w and w' and for any sequence z

$$|wz|_i \leq \|w\| \cdot |z|_i, \qquad |zw'|_i \leq |z|_i \|w'\| \tag{6.36}$$

provided that the products wz and zw' are well defined.

By Theorem 6.5, the left-hand side of (6.15) coincides with the norm $|\cdot|_0$ of the sequence on the left-hand side of (6.32). Hence applying (6.35), (6.36), definitions (6.14), and easy estimates

$$\|\mu H\| = \left\|\mu\Pi DR(I - DR)^{-1}\right\| \leq \frac{\|\mu DR\|}{1-\varepsilon},$$
$$\|\alpha G\| = \left\|\alpha R(I - DR)^{-1}\right\| \leq \frac{\|\alpha R\|}{1-\varepsilon}$$

for $\mu = \mu\Pi$, $\alpha \in \mathfrak{M}_0$ we conclude from (6.32) that the left-hand side of (6.15) does not exceed

$$\begin{aligned}
&|\alpha([Q] - [Q_0])f| \\
&\quad \leq \|\nu_0 - \nu\|(1 + |[Q]|_0)\|f\| + \left(|\sigma|_1 \left|\alpha[Q_0]^0 H\right|_1 + \|\alpha G\| |\delta|_1\right) 2\|D\Pi\| \, |[Q]|_0 \, \|f\| \\
&\quad = \|\nu_0 - \nu\|(1 + q)\|f\| + 2q\varepsilon_0\left(\sum_{t\geq 0} \left(\left\|\alpha\left(Q_0^t - \Pi_0\right)H\right\| + \|\alpha G\|\right)\|f\|\right) \\
&\quad \leq \frac{\varepsilon_{01}}{1-\varepsilon-\varepsilon r_0}(1 + r_0)(1 + q)\|f\| + 2q\frac{\varepsilon_0}{1-\varepsilon} r(\alpha)\|f\|.
\end{aligned} \tag{6.37}$$

We use here inequality (6.10) to estimate $\|\nu_0 - \nu\|$. Since $1 - \varepsilon - \varepsilon r_0 < 1 - \varepsilon$ and $1 + q \leq 2q$, (6.37) immediately implies (6.15). Theorem 6.3 is proved.

The inequality (6.17) of Theorem 6.4 is derived similarly from the representation (6.33) and from properties (6.35) and (6.36) for the introduced norms. □

To check the assertion of Remark 6.5 it is sufficient to observe that inequalities (6.15) and (6.16) used in above proofs remain valid if we calculate upper bounds and sums in definitions of norms $|\cdot|_i$ over the set $0 \leq t \leq t_0$, since the indicator function $\mathbb{I}_{\{t\leq t_0\}}$ possesses a semi-multiplicative property with respect to t in the domain $t \geq 0$. Making these replacements for the time domain we obtain a corresponding analogue of (6.15) and (6.17).

6.4. Uniform consolidation theorems

From estimates (6.15) and (6.17) one can derive approximations of distributions of a perturbed chain Y in a scheme of series that are uniform on classes of kernels Q and P and related characteristics. Note that the most strong conditions for the uniform ergodicity are posed only for the consolidated chain that is of simpler structure.

THEOREM 6.6. *Let conditions* (O), (I), (D), *and* (K) *be satisfied. Assume* $\alpha \in \mathfrak{M}_0$, $f \in \mathfrak{N}\mathcal{E}_0$. *If the consolidated chain* Y^0 *is uniformly ergodic and aperiodic so that*

$$\rho = \sup\left(\|\mu Q\Pi\|, \mu = \mu\Pi, \mu\mathbb{I} = 0, \|\mu\| = 1\right) < 1 \tag{6.38}$$

then the following asymptotic expansion is valid

$$\sup_{t\geq 0}\left|\mathsf{E}_\alpha f(Y_t) - \mathsf{E}_\alpha f\left(Y^0_y\right)\right| = O\left(\varepsilon_0 + \frac{\varepsilon_0\varepsilon}{1-\rho}\right), \tag{6.39}$$

provided that the right-hand side of (6.39) *tends to zero, where*

$$\varepsilon_0 = \|D\Pi\|, \qquad \varepsilon = \|DR\|. \tag{6.40}$$

The expansion (6.39) *is uniform with respect to* $\alpha \in \mathfrak{M}_0$ *with bounded* $\|\alpha\Pi\| + \|\alpha R\|$, *to* $f \in \mathfrak{N}\mathcal{E}_0$, *and to all basic chain and all kernels* Q *with bounded* $q + \|\Pi\| + \|\Pi_0\|$.

PROOF. The proof of (6.39) immediately follows from (6.15). Indeed, $S = \sum \|Q_0^t - \Pi_0\|$ is finite under condition (6.38) by Theorem 3.4 and, moreover,

$$S = O\left(\frac{1}{1-\rho}\right) \tag{6.41}$$

according to (3.14), since the norm $\|\Pi - \Pi_0\|$ is bounded. Then, the definition (6.14) implies that

$$r(\alpha) \leq \|\alpha R\| + \|\alpha\Pi\|S\|DR\| = O(\varepsilon S + 1).$$

Obvious estimates

$$\begin{gathered}\varepsilon_{01} = (\alpha\mathbb{I})\varepsilon_{01} = (\alpha\Pi\mathbb{I})\varepsilon_{01} \leq \|\alpha\Pi\|\cdot\|\mathbb{I}\|\cdot\|\nu_0 DR\| \leq \|\alpha\Pi\|\,\|\Pi_0\|\,\|R\| = O(\varepsilon),\\ r_0 = \|D\Pi R_0\| \leq \varepsilon_0\|R_0\|\end{gathered}$$

give us by Theorem 1.4 that

$$\varepsilon_{01}(1 + r_0) = O(\varepsilon(1 + \varepsilon_0 S)) = O(\varepsilon + \varepsilon\varepsilon_0 S),$$

where $\|R_0\| \leq S$. Therefore the numerator in (6.15) is

$$O(\varepsilon + \varepsilon_0 + \varepsilon\varepsilon_0 S) \tag{6.42}$$

uniformly on the class of α, f, P, and Q mentioned in the theorem. The definition of R_0 implies

$$\begin{gathered}\mu D\Pi R_0 = \mu(I - \Pi Q\Pi)R_0 = \mu(\Pi - \Pi_0),\\ 1 \leq \|\Pi - \Pi_0\| = \|D\Pi R_0\| \leq \|D\Pi\|\cdot\|R_0\| \leq \varepsilon_0 S\end{gathered}$$

for $\mu = \mu\Pi$; therefore the summand ε in (6.42) may be omitted, and from the convergence of $\varepsilon_0 + \varepsilon\varepsilon_0 S$ to zero it follows that $\varepsilon \to 0$. Therefore, $\varepsilon(1 + r_0) =$

$O(\varepsilon + \varepsilon\varepsilon_0 S)$ converges to zero. Thus the denominator in (6.15) is not less than $\frac{1}{2}$ for sufficiently small $\varepsilon + \varepsilon\varepsilon_0 S$.

Hence the right-hand side of (6.15) is

$$O(\varepsilon_0 + \varepsilon_0 \varepsilon S) \tag{6.43}$$

uniformly in the class of α, f, P, and Q mentioned above provided that (6.43) tends to zero.

Substituting the estimate (6.41) into (6.43) we obtain (6.39). The theorem is proved. □

Observe that the transition kernel of the consolidated chain Y^0 is equal to $Q_0 = \Pi + \Pi D \Pi$ by the definition and "almost" coincides with the projector. Thus distributions of Y_t^0 are approximated by distributions of a continuous time Markov process $Y^0(t)$ defined on $(E, \mathcal{E})$ by a semigroup of transition probabilities $Q_0(t) = \Pi \exp\{\Pi D \Pi t\}$. This semigroup is generated by the perturbation D averaged with the projector Π of the basic chain X.

Note also that invariant measures of the kernel Q_0 in (6.1) and of the semigroup $Q_0(t)$ coincide. According to (6.12), the stationary projector for $Q_0(t)$ is equal to $\Pi_0 = \mathbb{I} \circ \nu_0$. Besides, the sum $\Pi + \Pi D \Pi = Q_0$ is a stochastic kernel.

To estimate the closeness of distributions of Y_t^0 and $Y^0(t)$ we additionally assume that the semigroup $Q_0(t)$ is uniformly ergodic, so that

$$\omega(D\Pi) = \inf\left(\|\mu D \Pi\|, \mu = \mu\Pi, \mu\mathbb{I} = 0, \|\mu\| = 1\right) > 0. \tag{6.44}$$

Ergodic properties of such semigroups are described in the following lemma.

LEMMA 6.5. *Let $\Pi \in \mathcal{B}$ be a projector and a semigroup $Q_0(t) = \Pi \exp\{D\Pi t\}$ with the stationary projector Π_0 satisfy (6.44). Assume also that the kernel $\Pi + \Pi D \Pi \in \mathcal{B}$ is nonnegative and $\varepsilon_0 = \|D\Pi\|$. Then indices*

$$\begin{gathered} \sigma = \int_0^\infty \|Q_0(t) - \Pi_0\| \, dt, \\ p = \sup_{t \geq 0} \|Q_0(t) - \Pi_0\| \leq 3\varepsilon_0 \sigma \end{gathered} \tag{6.45}$$

are finite and

$$\|Q_0(t) - \Pi_0\| < 3\sigma\varepsilon_0 \exp\left\{-\frac{t}{2\sigma}\right\}. \tag{6.46}$$

PROOF. First we prove (6.46). Consider the resolvent of the semigroup $(\Pi - \Pi_0)Q_0(t) = Q_0(t) - \Pi_0$:

$$D_z = (\Pi - \Pi_0)(z - D\Pi)^{-1} = \int_0^\infty \exp\{-zt\}(Q_0(t) - \Pi) \, dt. \tag{6.47}$$

The spectrum of $D\Pi$, as an operator on $\mathfrak{M}\Pi$, belongs to the domain $\Re(z) \leq 0$. Moreover, for $|z| \geq 2\varepsilon_0$,

$$\|D_z\| \leq \|\Pi - \Pi_0\|(|z| - \|D\Pi\|)^{-1} \leq \frac{\|\Pi - \Pi_0\|}{\varepsilon_0}. \tag{6.48}$$

Thus, D_z is analytical for $|z| \geq 2\varepsilon_0$ and $\Re(z) > 0$.

Let us introduce the index

$$\sigma_0 = \sup\left(\|D_z\|, \Re(z) = 0, |z| < 2\varepsilon_0\right). \tag{6.49}$$

We show that (6.46) is valid for $\sigma = \sigma_0$. Since the right-hand side of (6.46) is monotone with respect to σ and (6.47) implies $\sigma_0 \leq \int_0^\infty \|Q_0(t) - \Pi_0\|\, dt = \sigma$, (6.46) is proved.

Take $\delta = 1/2\sigma_0$ and consider the contour Γ in the complex plane formed by the interval Γ^1 of the straight line $\Re(z) = -\delta$ inside the circle $|z| = 2\varepsilon_0$ and by the left arc Γ^2 of this circle. The spectrum of $Q_0(t) - \Pi_0$ lies inside Γ, since, for $\Re(z) = 0$, from the resolvent identity $D_{z-\delta} = D_z + \delta D_z D_{z-\delta}$ follows that

$$\sup\left(\|D_{z-\delta}\|, \Re(z) = 0, |z| < 2\varepsilon_0\right) \leq 2\sigma_0. \tag{6.50}$$

Using estimate (6.50) for $z \in \Gamma^1$ and estimate (6.48) for $z \in \Gamma^2$ we obtain

$$\begin{aligned} \|Q_0(t) - \Pi_0\| &= \left\| \frac{1}{2\pi i} \int_\Gamma \exp\{zt\} D_z\, dz \right\| \\ &\leq \frac{1}{2\pi}\left(\pi 2\varepsilon_0 \|\Pi - \Pi_0\| \varepsilon_0^{-1} + 4\varepsilon_0 \cdot 2\sigma_0\right) \sup_\Gamma |\exp\{zt\}| \\ &\leq 3\varepsilon_0\sigma_0 \left\{-\frac{t}{2\sigma_0}\right\}, \end{aligned}$$

since $\|\Pi - \Pi_0\| \leq \varepsilon_0\sigma_0$. Inequality (6.46) is proved.

The index σ_0 is finite because of the estimate (6.43). To prove that σ is finite we integrate inequality (6.46) for $\sigma = \sigma_0$ and obtain $\sigma \leq 6\varepsilon_0\sigma_0^2 < \infty$. The lemma is proved. □

Note that σ_0 may be estimated also by Theorem 3.1 applied to the kernel $\Pi + \Pi D \Pi$.

THEOREM 6.7. *Let conditions* (O), (I), (D), (K), *and* (6.44) *be satisfied. Assume $f \in \mathfrak{N}\mathcal{E}_0$ and that an initial distribution α belongs to $\mathfrak{M}_0$. Then the following asymptotic expansion is valid*

$$\sup_{t \geq 0} \left|\mathsf{E}_\alpha f(Y_t) - \mathsf{E}_\alpha f\left(Y^0(t)\right)\right| = O\left((\varepsilon_0 + \varepsilon)\sigma\varepsilon_0 p\left(1 + \ln^+\left(\frac{\sigma\varepsilon_0}{p}\right)\right)\right) \tag{6.51}$$

provided that the right-hand side containing indices (6.40) *and* (6.45) *converges to zero. Expansion* (6.51) *is uniform with respect to $\alpha \in \mathfrak{M}_0$ with bounded $\|\alpha\Pi\| + \|\alpha R\|$, to $f \in \mathfrak{N}\mathcal{E}_0$ with $\|f\| \leq 1$, to all perturbations D, and to all basic chains X such that $q + \|\Pi_0\| + \|\Pi\|$ is bounded.*

REMARK 6.6. It follows from (6.45) that the right-hand side of (6.51) tends to zero if $\sigma = o\left((\varepsilon_0 + \varepsilon)^{-3/2}\right)$. Here σ is negatively homogeneous so that the product $\varepsilon_0\sigma$ does not depend on the scale factor of D. Thus, in particular, the right-hand side of (6.45) is $Q(\delta)$ as $\delta \to 0$ if the basic chain X is not changing in the scheme of series and the perturbation is of the form $D = \delta B$, where the kernel B is fixed and satisfies conditions (I), (D), and (6.44) instead of D.

PROOF. Consider the stochastic kernel $\overline{Q} = Q_0(1) = \Pi \exp\{\Pi D \Pi\}$. It is clear that $\overline{Q}^t = Q_0(t)$ for $t \geq 0$. So the difference on the left-hand side of (6.45) is equal to

$$\alpha\left(Q^t - Q_0(t)\right) f = \alpha\left(Q^t - Q_0^t\right) f + \alpha\left(Q_0^t - \overline{Q}^t\right) f. \tag{6.52}$$

The first summand in (6.52) was estimated in the proof of Theorem 6.6 by expression (6.43), where $S = \sum_{t\geq 0} \|Q_0^t - \Pi_0\|$. Since

$$\left\|Q_0 - \overline{Q}\right\| = \|\Pi + \Pi D \Pi - \Pi \exp\{\Pi D \Pi\}\| = O(\|\Pi D \Pi\|^2) = O\left(\varepsilon_0^2\right) \tag{6.53}$$

because $\|\Pi\|$ is bounded, we get in the same manner as in the proof of Theorem 6.3 that

$$\begin{aligned} \left\|Q_0^t - \overline{Q}^t\right\| &= \left\|\left(Q_0^t - \overline{Q}^t\right)(\Pi - \Pi_0)\right\| \\ &= \left\|\sum_{s<t}\left(\overline{Q}^{t-1-s} - \Pi_0\right)\left(\overline{Q} - Q_0\right)\left(Q_0^s - \Pi_0\right)\right\| \\ &\leq \sup_{t\geq 0}\left\|\overline{Q}^t - \Pi_0\right\| \|\overline{Q} - Q_0\| \sum_{s\geq 0} \|Q_0^s - \Pi_0\| = O(p\varepsilon_0^2 S) \end{aligned} \tag{6.54}$$

uniformly in t. Here we use definition (6.45) of the constant p.

It follows from (6.52), (6.43), and (6.54) that the left-hand side of (6.51) is

$$O(\varepsilon_0 + \varepsilon\varepsilon_0 S + p\varepsilon_0^2 S) = O((\varepsilon + \varepsilon_0)p\varepsilon_0 S) \tag{6.55}$$

uniformly in those α, f, P, and Q that are described in the statement of the theorem. The equality in (6.55) is valid, since the values of p and $\varepsilon_0 S \geq 1$ are isolated from zero (see the proof of Theorem 6.6).

It remains to estimate S in (6.55). Set $\overline{S} = \sum \left\|\overline{Q}^t - \Pi_0\right\|$. Applying a similar to (6.54) inequality we obtain

$$\begin{aligned} S &\leq \sum_{t\geq 0} \left\|Q_0^t - \overline{Q}^t\right\| + \sum_{t\geq 0} \left\|\overline{Q}^t - \Pi_0\right\| \\ &\leq \sum_{t\geq 0} \left\|\sum_{s<t}\left(\overline{Q}^{t-1-s} - \Pi_0\right)\left(\overline{Q} - Q_0\right)\left(Q_0^s - \pi_0\right)\right\| + \overline{S} \\ &\leq \overline{S}\varepsilon_0^2 \|\Pi\|^2 S + \overline{S}. \end{aligned} \tag{6.56}$$

It follows from (6.56) that $S = O(\overline{S})$ provided $\varepsilon_0^2 \overline{S} \to 0$. To estimate $\overline{S}$ we apply inequality (6.46) to the semigroup $Q_0(t)$

$$\left\|\overline{Q}^t - \Pi_0\right\| = \|Q_0(t) - \Pi_0\| \leq \min\left(p, 3\varepsilon_0\sigma \exp\left\{-\frac{t}{2\sigma}\right\}\right).$$

Summing up these inequalities over $t \geq 0$ we obtain

$$\begin{aligned}\overline{S} &= \sum_{t\geq 0} \|Q_0^t - \Pi_0\| \leq 3\sum_{t\geq 0} \min\left(p, \varepsilon_0\sigma \exp\left\{-\frac{t}{2\sigma}\right\}\right) \\ &= 6p\sigma \ln^+ \frac{\varepsilon_0\sigma}{p} + 3p\left(1 - \exp\left\{-\frac{1}{2\sigma}\right\}\right) = O\left(p\sigma\left(1 + \ln^+ \frac{\varepsilon_0\sigma}{p}\right)\right)\end{aligned} \tag{6.57}$$

uniformly in p and σ.

From (6.57) and from the convergence to zero of the right-hand side of (6.51) it follows that $\varepsilon_0^2\overline{S} \to 0$, so $S = O(\overline{S})$ as was shown above. The last expression is estimated by (6.57). Substituting this estimate into (6.55) we arrive at (6.51). The theorem is proved. □

The theorems above describe the uniform in $t \geq 0$ approximation of distributions of Y. Using Remark 6.5 we obtain in (6.44) an estimate of such an approximation on a finite interval of t in terms of a characteristic ω that is simpler than σ. The evaluation of ω may be reduced to determining the inverse operator $(\Pi - \Pi_0)(\Pi D \Pi)^{-1} = R_0$, since $\|R_0\| = \omega^{-1}$. To simplify the statements we restrict ourselves to the case where $\mathfrak{M}$ consists of all finite measures with the total variation norm.

THEOREM 6.8. *Let $\|\mu\| = |\mu|(E)$ and conditions* (O), (I), (D), *and* (6.44) *be satisfied. Assume that $\alpha \in \mathfrak{M}_0$. Then the following asymptotic expansion is valid*

$$\sup_{t\leq t_0} \sup_{A\in\mathcal{E}_0} \left|\mathsf{P}_\alpha(Y_t \in A) - \mathsf{P}_\alpha\left(Y_t^0 \in A\right)\right| = O\left(\varepsilon_0 + \varepsilon_0\left(t_0 + +\frac{1}{\omega}\right)\right) \tag{6.58}$$

in any domain $t_0 = o(1/\varepsilon\varepsilon_0)$, where the constant $\omega = \omega(D\Pi)$ is given by (6.44). *Moreover, expansion* (6.58) *is uniform in $\alpha \in \mathfrak{M}_0$ with bounded $\|\alpha R\|$, in all perturbations D, and in all basic chains X, so that the expression $O(\cdot)$ is determined by an absolute constant.*

PROOF. First, note that operators P, Q, Q_0, Π, and Π_0 are contractions on $\mathfrak{M}$. Using Remark 6.5 we estimate in (6.15) $r(\alpha) \leq \|\alpha R\| + 2\varepsilon t_0$, $\varepsilon_{01} \leq \varepsilon$, $q = 1$, and $r_0 = \|DR_0\| = \|D\Pi R_0\| \leq \varepsilon_0\|R_0\|$. Here $\mu = \alpha R_0$ is equivalent to $\mu = \mu\Pi$, $\mu\mathbb{I} = 0$, and $\mu D\Pi = \alpha(\Pi - \Pi_0)$ in view of the definition of R_0. Then from (6.44) we derive that $\|\alpha R_0\| = \|\mu\| \leq \|\alpha\|/\omega$, so that $\|R_0\| \leq \omega^{-1}$. Thus replacing the set $t \geq 0$ by $0 \leq t \leq t_0$ in the definition of the upper bound in (6.15), we transform the right-hand side of (6.15) to the form

$$O\left(\varepsilon_0 + \varepsilon\varepsilon_0 t_0 + \varepsilon + \varepsilon\frac{\varepsilon_0}{\omega}\right),$$

where $\varepsilon = O(\varepsilon\varepsilon_0/\omega)$, since $\omega \leq \varepsilon_0$ and $\varepsilon\varepsilon_0 t_0 \to 0$. Substituting these relations into the expression above we arrive at (6.58). The theorem is proved. □

CHAPTER VII

ESTIMATES OF THE EXPONENTIAL ASYMPTOTIC OF MARKOV MOMENTS

In this Chapter, we investigate the asymptotics of Markov moments τ on the basic chain X taking values in $(E, \mathcal{E})$. The distribution of the moment is given, unlike for the schemes of Chapter 6, by a strictly substochastic kernel subordinated to the transition kernel of the chain X. These kernels are close to each other, so τ may be treated as a killing moment.

We consider the scheme of series where the finite mean of the moment τ tends to infinity. In this case, the distribution of τ is approximated by a geometric distribution, and the last is approximated by an exponential distribution. Below, we estimate the accuracy of such a kind of approximation and give explicit inequalities.

Estimates mentioned above may be derived from results of Chapter 6 by an appropriate construction of the extended Markov chain and of the consolidated chain with two states. As a result of this approach we obtain an accompanying distribution with the parameter that is equal to the probability of the event $\{\tau = 1\}$. The approximation accuracy is of the first order with respect to the norm of the perturbation.

Below we use an another method. We introduce a geometric distribution with a different parameter that coincides with the reciprocal to the expectation of τ. For this case, the approximation accuracy may be improved to the second order with respect to the norm of the perturbation.

Throughout we shall assume the condition (M1) for the space $\mathfrak{M}$. Also either (M2) or (M2′) is assumed.

7.1. Basic representations for the distribution of τ

Consider a basic homogeneous Markov chain $X = (X_t, t \geq 0)$ taking values in a measurable space $(E, \mathcal{E})$. This chain is given by the transition kernel $P(x, A)$ and has a unique invariant stochastic measure π. By P_x and E_x we denote the conditional probability and expectation on the initial probability space $(\Omega, \mathfrak{F}, \mathsf{P})$ under the condition $X_0 = x$. The symbols P_α and E_α have the same meaning for an initial distribution α on $\mathcal{E}$, so that $\mathsf{P}_x(X_1 \in A) = P(x, A)$, $\mathsf{P}_\alpha(X_1 \in A) = \alpha \mathsf{P}(A)$.

Assume that the chain X is adapted to a flow $(\mathfrak{F}_t, t \geq 0)$ of σ-fields on the space $(\Omega, \mathfrak{F}, \mathsf{P})$. Let τ be a homogeneous killing moment with respect to the system $(\Omega, \mathfrak{F}, \mathfrak{F}_t, \mathsf{P})$, that is,

$$\mathsf{P}_x\left(\tau > t+1, X_{t+1} \in A / \mathfrak{F}_t\right) = \mathsf{P}_{X_t}(\tau > 1, X_1 \in A) \mathbb{I}_{\{\tau > t\}} \tag{7.1}$$

P_x-almost surely for all $t \geq 0$, $A \in \mathcal{E}$, $x \in E$.

In particular, one can take a moment

$$\tau_C = \inf\left(t \geq 1 \colon (X_{t-1}, X_t, \xi_t) \in C\right) \tag{7.2}$$

as τ, where the set C is from $\mathcal{E} \times \mathcal{E} \times \mathcal{B}(\mathbb{R})$ and random variables ξ_t are independent identically distributed and do not depend on X. The flow $\{\mathfrak{F}_t\}$ is defined in this case as follows

$$\mathfrak{F}_t = \sigma\left[X_s, \xi_s, s \leq t\right].$$

The distribution of τ is uniquely determined by the substochastic kernel

$$Q(x, A) = \mathsf{P}_x(\tau > 1, X_1 \in A), \tag{7.3}$$

since (7.1) implies that

$$\mathsf{P}_x(\tau > t, X_t \in A) = Q^t(x, A), \tag{7.4}$$

where the right-hand side contains the power of the kernel Q.

Note that, for all $x \in E$, $A \in \mathcal{E}$, the kernel Q is subordinated to the kernel P, that is, $Q(x, A) \le P(x, A)$. Thus the difference

$$D(x, A) = P(x, A) - Q(x, A) = \mathsf{P}_x(\tau = 1, X_1 \in A)$$

is nonnegative.

To prove (7.4) we show that

$$\mathsf{E}_x f(X_t)\mathbb{I}_{\{\tau > t\}} = \int_E Q^t(x, dy)\, f(y) = Q^t f(x)$$

for all $t \ge 0$ and all measurable bounded functions f. For $t = 1$, this equality follows from (7.3). Next, if it is valid for a given t then

$$\begin{aligned} \mathsf{E}_x \left(f(X_{t+1})\mathbb{I}_{\{\tau > t+1\}}\right) &= \mathsf{E}_x \left(\mathsf{E}_x \left[f(X_{t+1})\mathbb{I}_{\{\tau > t+1\}}/\mathfrak{F}_t\right]\right) \\ &= \mathsf{E}_x \left(\mathbb{I}_{\{\tau > t\}}\, \mathsf{E}_x \left[f(X_{t+1})\mathbb{I}_{\{\tau > t+1\}}/\mathfrak{F}_t\right]\right) \\ &= \mathsf{E}_x \left(\mathbb{I}_{\tau > t}\, \mathsf{E}_{x_t}(f(X_1)\mathbb{I}_{\tau > 1})\right) \\ &= \mathsf{E}_x \left(\mathbb{I}_{\{\tau > t\}} Q f(X_t)\right) = Q^t(Qf)(x) = Q^{t+1} f(x) \end{aligned}$$

by the definition of the power of a transition kernel Q.

Suppose also that the mean $m = \mathsf{E}_\pi \tau$ is finite and varies in the scheme of series, so that $m \to \infty$. Then $D(x, E) = \mathsf{P}_x(\tau = 1) \to 0$ in average by the measure π, that is, the kernel $Q = P - D$ approaches P. The converse is also true: If $D \to 0$ in a topology defined below then $m \to \infty$, $\tau \to \infty$, and the distribution of τ approaches an exponential distribution. We shall describe the convergence of D to zero in terms of the norm $\|\cdot\|$ on $\mathfrak{M}$ and $\mathcal{B}$ that was introduced in Chapter 1.

In the sequel the quantity $\rho = 1 - m^{-1}$ stands for the parameter of a geometric distribution on $\mathbb{N}$ with the mean m.

Introduce $\mathcal{B}$-valued sequences $[Q]$ and $[P]$ with members $[Q]_t = Q^t$ and $[P]_t = P^t$ respectively. Let also numerical sequences G, g, σ, and δ be given by

$$G_t = \rho^t, \qquad g_t = (1 - \rho)\rho^{t-1}, \qquad g_0 = 0, \qquad \sigma_t = \delta_{1t}, \qquad \delta_t = \delta_{0t}.$$

According to (7.4)

$$(\alpha[Q]f)_t = \alpha Q^t f = \mathsf{E}_\alpha \left(f(X_t)\mathbb{I}_{\{\tau > t\}}\right).$$

Operations with sequences such as multiplication by a number, by a measure, by a function, and also as a convolution are understood in the same sense as above in proofs of Theorems 6.3 and 6.4. Below we use the norms $|\cdot|_i$ and their properties (6.35) and (6.36).

Recall that the inequalities $\mu \le \nu$ for measures μ, ν and $f \le g$ for functions f, g mean that $\mu(A) \le \nu(A)$ for all $A \in \mathcal{E}$ and $f(x) \le g(x)$ for all $x \in E$, respectively.

We use in what follows the following auxiliary result.

LEMMA 7.1. *For all $t \geq 1$, the following inequalities are valid*

$$\pi Q^{t+1} \leq \pi Q^t, \qquad Q^{t+1}\mathbb{I} \leq Q^t\mathbb{I}. \tag{7.5}$$

In particular, $\pi Q^t \prec\!\prec \pi$.

PROOF. The kernels $D = P - Q$ and Q are nonnegative and the kernel P has the stochastic property. Thus (7.5) is valid, since

$$\pi Q^t - \pi Q^{t+1} = (\pi - \pi Q)Q^t = (\pi P - \pi Q)Q^t = \pi D Q^t \geq 0,$$
$$Q^t\mathbb{I} - Q^{t+1}\mathbb{I} = Q^t(\mathbb{I} - Q\mathbb{I}) = Q^t(P\mathbb{I} - Q\mathbb{I}) = Q^t D\mathbb{I} \geq 0.$$

□

Consider the following equations in the spaces of measures $\nu \in \mathfrak{M}$ and functions $h \in \mathfrak{N}$:

$$\nu = \nu Q + \pi(\rho I - Q), \tag{7.6}$$

$$h = Qh + (\rho I - Q)h \tag{7.7}$$

under assumption that spaces $\mathfrak{M}$ and $\mathfrak{N}$ are invariant for Q.

THEOREM 7.1. *Let $\nu \in \mathfrak{M}$ and $h \in \mathfrak{N}$ satisfy (7.6) and (7.7) in $\mathfrak{M}$ and $\mathfrak{N}$, respectively. Then, respectively, the following representations are valid*

$$\pi[Q] - G\pi = -G\nu + \nu[Q] * (\delta - g), \tag{7.8}$$

$$[Q]\mathbb{I} - G\mathbb{I} = -Gh + (\delta - g) * [Q]\mathbb{I}. \tag{7.9}$$

PROOF. We check (7.8) under condition (7.6). For $t = 0$, the left and the right-hand sides of (7.8) vanish. For $t \geq 1$, we obtain just from the definition of the convolution

$$\begin{aligned}
\big(\nu[Q] * (\delta - g)\big)_t &= \nu Q^t - \sum_{s=0}^{t-1} \nu Q^s (1-\rho)\rho^{t-1-s} \\
&= \nu Q^t - \sum_{s=1}^{t} \nu Q^{s-1}\rho^{t-s} + \sum_{s=0}^{t-1} \nu Q^s \rho^{t-s} \\
&= -\nu(I - Q)Q^{t-1} + \nu\rho^t - \sum_{s=1}^{t-1} \nu(I - Q)Q^{s-1}\rho^{t-s} \\
&= \nu\rho^t - \sum_{s=1}^{t-1} \nu(I - Q)Q^{s-1}\rho^{t-s} \\
&= \nu\rho^t - \sum_{s=0}^{t-1} \pi Q^s\rho^{t-s} + \sum_{s=1}^{t-1} \pi Q^s \rho^{t-s} \\
&= (\nu)\rho^t - (\pi)\rho^t + \pi Q^t = \big(\pi[Q] - (\pi)G + (\nu)G\big)_t,
\end{aligned}$$

which proves (7.8). The representation (7.9) may be proved similarly. □

7.2. Approximations of distributions of "rare" Markov moments

We obtain first an estimate of the asymptotics of τ in terms of expectations with respect to measures P_x.

THEOREM 7.2. *Let $\mathsf{E}_\pi \tau < \infty$, so that the function $m(x) = \mathsf{E}_x \tau$ is integrable with respect to the measure π. Assume that the initial distribution α is absolutely continuous with respect to π and $\alpha(A) \le a\pi(A)$ for all $A \in \mathcal{E}$. Then, for any fixed $\rho = 1 - 1/m$, and in particular for $m = \mathsf{E}_\pi \tau$,*

$$\sup_{t\ge 0} \left|\mathsf{P}_\alpha(\tau > t) - \rho^t\right| \le 2\frac{a}{m} \int \pi(dx)\, |m(x) - m|. \tag{7.10}$$

PROOF. Define a function $h(x) = 1 - m(x)/m$. It is easy to obtain that

$$\begin{aligned} Qm(x) &= \mathsf{E}_x(m(X_1), \tau > 1) = \mathsf{E}_x(E_{X_1}(\tau, \tau > 1)) = \mathsf{E}_x\left(\mathsf{E}_{X_1}\left[(\tau-1)/\mathfrak{F}_t\right] \mathbb{I}_{\{\tau>1\}}\right) \\ &= \mathsf{E}_x\left((\tau - 1)\mathbb{I}_{\{\tau>1\}}\right) = \mathsf{E}_x(\tau - 1) = m(x) - 1. \end{aligned}$$

These equalities follow from definition (7.1) of the killing moment τ. Hence

$$h(x) - Qh(x) = 1 - \frac{m(x)}{m} - Q\mathbb{I}(x) + \frac{m(x)-1}{m} = 1 - \frac{1}{m} - Q\mathbb{I}(x) = (\rho I - Q)\mathbb{I}(x)$$

provided that $m(x) < \infty$. Since $\mathsf{E}_\pi \tau < \infty$ by the condition, $m(x) < \infty$ π-almost surely. Thus the function h satisfies (7.7) for all measures from $\{\alpha : \alpha \le a\pi\}$. The last set is Q-invariant by Lemma 7.1. Thus representation (7.9) holds true for the distribution α, whence using (6.35) we obtain

$$\begin{aligned} \left|\alpha[Q]\mathbb{I} - G\right|_0 &= |-G(\alpha h) + (\delta - g) * \alpha[Q]h|_0 \\ &\le |\alpha[Q]h|_0 + |G(\alpha h) + g * \alpha[Q]h|_0 \\ &\le 2|\alpha[Q]h|_0 \le 2\sup_{t\ge 0}\left(\alpha Q^t|h|\right) \le 2a \sup_{t\ge 0} \pi Q^t |h| \le 2a\pi|h| \\ &= 2a \int \pi(dx) \left|1 - \frac{m(x)}{m}\right| \end{aligned} \tag{7.11}$$

that proves (7.10). □

COROLLARY 7.1. *Let a chain X have a regeneration moment ξ, that is, σ-fields $\mathfrak{F}_\xi$ and $\sigma[X_{t+\xi}, t \ge 0]$ are independent given X_ξ. Let $X_\xi = \overline{x}$ be nonrandom P_x-almost surely for all $x \in E$, $\mathsf{E}_\pi \xi < \infty$, and $\alpha(A) \le a\pi(A)$ for all $A \in \mathcal{E}$. Then*

$$\sup_{t\ge 0} \left|\mathsf{P}_\alpha(\tau > t) - \left(1 - \frac{1}{\overline{m}}\right)^t\right| \le 2a\left(\mathsf{P}_\pi(\tau < \xi) + \frac{\mathsf{E}_\pi \xi}{\overline{m}}\right), \tag{7.12}$$

where $\overline{m} = \mathsf{E}_{\overline{x}} \tau$.

PROOF. The proof of the corollary follows from the evident representation

$$\begin{aligned} m(x) &= \mathsf{E}_x(\tau, \tau < \xi) + \mathsf{E}_x(\tau, \tau \ge \xi) = \mathsf{E}_x(\tau, \tau < \xi) + \mathsf{E}_x(\xi + \mathsf{E}_{\overline{x}}\tau, \tau \ge \xi) \\ &= \overline{m}\,(1 - \mathsf{P}_x(\tau < \xi)) + \mathsf{E}_x \min(\tau, \xi). \end{aligned} \tag{7.13}$$

Obviously (7.13) implies the finiteness of $\mathsf{E}_\pi\,\tau$. Therefore (7.12) follows from (7.10) with $m = \overline{m}$. In so doing, we evaluate $m(x) - \overline{m}$ from (7.13) and estimate $\min(\tau, \xi) \le \xi$. The corollary is proved. □

Inequality (7.10) is applicable when the mean $m(x)$ is known. Let us estimate the asymptotics of τ in terms of a "small" kernel $D = P - Q$ and the potential of the basic chain X. In the sequel the parameter m always means the average $m = \mathsf{E}_\pi\,\tau$ and, correspondingly,

$$\rho = 1 - \frac{1}{m} = 1 - \frac{1}{\mathsf{E}_\pi\,\tau},$$

and $p = \mathsf{P}_\pi(\tau = 1) = \pi D \mathbb{I}$.

Denote, as in Section 1.2, by $\Pi = \mathbb{I} \circ \pi$ the stationary projector of the kernel P and let, according to Definition 1.3, the subspace $\mathfrak{M}_0 = \mathfrak{M}(I - P + \Pi)$ be the domain of the potential R of the chain X. As in Chapter 6, we also introduce the subspace $\mathfrak{N}_0 = (I - P + \Pi)\mathfrak{N}$ and define, on $\mathfrak{N}_0$, values of the potential R as follows

$$R(I - P) = I - \Pi, \qquad R\Pi = 0.$$

To be correct, we suppose additionally that the kernel P has no invariant functions except constants in $\mathfrak{N}_0$. In this case the operator $I - P + \Pi$ has no annihilators in $\mathfrak{N}$. Observe that the introduced condition is equivalent to that $\mathfrak{M}_0$ is dense in $\mathfrak{M}$, [see Rudin (1973), Theorem 4.12].

As it is seen from Theorem 5.2, for general chains X that are not uniform ergodic, one cannot expect the appropriate properties of the asymptotics of the moment τ if the kernel D is small only in the operator topology. The class of admittable perturbations $D = P - Q$ is described in terms of the range $\mathfrak{M}D$. For easy statements of dual results we assume that the kernel D admits a factorization

$$D = \Delta'\Delta, \qquad \Delta \in \mathcal{B}^+,\ \Delta' \in \mathcal{B}^+, \tag{7.14}$$

where one of kernels Δ, Δ' equals to the identity. Suppose also that one of the conditions below is fulfilled:

(MD) $\mathfrak{M}\Delta \subset \mathfrak{M}_0$ and the dual norm on $\mathfrak{N}$ satisfies condition (M2′);
(MD′) $\Delta'\mathfrak{N} \subset \mathfrak{N}$, $\mathfrak{M}_0$ is dense in $\mathfrak{M}$, and the norm on $\mathfrak{M}$ satisfies condition (M2).

If either (MD) or (MD′) is satisfied then, by Lemma 5.1 and by its evident dual analogue, the product operator $\Delta R \Delta'$ is well defined and bounded on $\mathfrak{M}$. Besides, (MD) implies that ΔR and DR are bounded and (MD′) implies that $R\Delta'$ and RD are bounded. Thus, under appropriate assumptions, the following indices are well defined

$$\begin{aligned} \varepsilon &= \|\Delta R \Delta'\|, & \kappa &= \|\mathbb{I}\|, \varkappa = \|\pi\|, \\ \varepsilon_0 &= \|\pi\Delta'\| \cdot \|\Delta R\|, & \varepsilon_0' &= \|R\Delta'\| \cdot \|\Delta \mathbb{I}\|, \end{aligned} \tag{7.15}$$

where $\mathbb{I} \in \mathfrak{N}$ is the function that identically equals unity.

In some cases we assume that

$$q = \sup_{t \ge 0} \|Q^t\|\,. \tag{7.16}$$

is finite. The index q is finite if, in particular, X is uniformly ergodic and aperiodic for all sufficiently small $\|D\|$.

LEMMA 7.2. *Let either condition* (MD) *or* (MD′) *be satisfied and let* $\varepsilon < 1$. *Then the mean value* $m = \mathsf{E}_\pi \tau$ *is finite and admits the following representation*

$$\frac{1}{m} = \pi\Delta'(I + \Delta R\Delta')^{-1}\Delta\mathbb{1}. \tag{7.17}$$

Moreover, the measure $\nu \in \mathfrak{N}$ *and the function* $h \in \mathfrak{N}$ *that are defined in conditions* (MD) *and* (MD′) *by equalities*

$$\begin{aligned} \nu &= \pi\Delta'\left(I + \Delta R\Delta'\right)^{-1}\Delta R, \\ h &= R\Delta'\left(I + \Delta R\Delta'\right)^{-1}\Delta\mathbb{1}, \end{aligned} \tag{7.18}$$

respectively, satisfy equalities (7.6) *and* (7.7) *in the spaces* $\mathfrak{M}$ *and* $\mathfrak{N}$ *respectively.*

PROOF. Under conditions of the lemma $(I + \Delta R\Delta')^{-1} \in \mathcal{B}$ and $\pi \in \mathfrak{M}$, $\mathbb{1} \in \mathfrak{N}$. Let condition (MD) be satisfied and $\varepsilon < 1$. Then evidently $\nu \in \mathfrak{M}$ and $DR \in \mathcal{B}$.
By (7.18) and (7.14) we calculate

$$\begin{aligned} (\pi - \nu)(I + DR) &= \pi(I + DR) - \pi\Delta'(I + \Delta R\Delta')^{-1}\Delta R(I + \Delta'\Delta R) \\ &= \pi(I + \Delta'\Delta R) - \pi\Delta'(I + \Delta R\Delta')^{-1}(I + \Delta R\Delta')\Delta R \\ &= \pi(I + \Delta' R\Delta) - \pi\Delta'\Delta R = \pi. \end{aligned}$$

Thus by Definition 1.3 of the potential R we get

$$0 = \pi(I - P) = (\pi - \nu)(I - P + D)(I - \Pi)) = (\pi - \nu)(I - Q - D\Pi).$$

Since the measure $(\pi - \nu)D\Pi = (\pi - \nu)D\mathbb{1}\pi$ is proportional to π, we derive

$$\nu(I - Q) = \pi(rI - Q), \tag{7.19}$$

where $r = 1 - (\pi - \nu)D\mathbb{1}$ is a constant. Therefore, it is sufficient to establish that m is finite and that $r = 1 - m^{-1} = \rho$. In this case (7.17) follows from (7.18), since

$$\begin{aligned} \frac{1}{m} &= 1 - (1 - (\pi - \nu)D\mathbb{1}) = \pi\Delta'\Delta\mathbb{1} - \pi\Delta'(I + \Delta R\Delta')^{-1}\Delta\mathbb{1} \\ &= \pi\Delta'\left(I - (I + \Delta R\Delta')^{-1}\Delta R\Delta'\right)\Delta\mathbb{1} = \pi\Delta'(I + \Delta R\Delta')^{-1}\Delta\mathbb{1}. \end{aligned}$$

To evaluate r we iterate equation (7.19) t times:

$$\begin{aligned} \nu &= \nu Q^t + r\pi\left(I + \cdots + Q^{t-1}\right) - \pi\left(I + \cdots + Q^t\right) \\ &= (\nu - \pi)Q^t + \pi - (1 - r)\pi\left(I + \cdots + Q^{t-1}\right). \end{aligned} \tag{7.20}$$

Assumptions of the lemma imply the inclusion $\nu \in \mathfrak{M}$ and thus the measure ν is finite. Since Q is substochastic, the measures νQ^t are also uniformly bounded. Therefore, the sequence

$$\nu_t(A) = \sum_{s<t} \pi Q^s(A)$$

is monotone for all $A \in \mathcal{E}$ and is uniformly bounded with respect to t and A. Thus there exists the limit $\nu_0(A) = \lim_{t\to\infty} \nu_t(A)$ and this limit is a finite measure on $\mathcal{E}$. Since

$$\nu_0(E) = \sum_{s\geq 0} \pi Q^s(E) = \sum_{s\geq 0} \mathsf{P}_\pi(\tau > s) = \mathsf{E}_\pi \tau = m,$$

we obtain $m < \infty$.

Using the convergence of ν_t we conclude from (7.20) that, for all $A \in \mathcal{E}$, there exists the monotone limit $\lim_{t\to\infty}(\pi-\nu)Q^t(A)$ that coincides with the finite measure $\overline{\nu} = \pi - \nu - (1-r)\nu_0$. Then the measure $\overline{\nu}$ is invariant for Q and from $\overline{\nu} = \overline{\nu}Q$ we derive that $|\overline{\nu}| \leq |\overline{\nu}|Q \leq |\overline{\nu}|\mathsf{P}$. But $|\overline{\nu}|\mathbb{I} = |\overline{\nu}|\mathsf{P}\mathbb{I}$. Thus $|\overline{\nu}|(A) = |\overline{\nu}|\mathsf{P}(A)$ for all $A \in \mathcal{E}$. Hence $|\overline{\nu}| = a\pi$ because the measure π is unique. If $a \neq 0$ then $\pi = \pi Q = \pi\mathsf{P}$ that contradicts the finiteness of the measure ν_0 introduced above. Thus $a = 0$ and $\pi - \nu = (1-r)\nu_0$. But from definition (7.18) it follows that $\nu\mathbb{I} = 0$ and $\pi\mathbb{I} = 1$ by the definition of π. Then $1 = \pi\mathbb{I} - \nu\mathbb{I} = (1-r)\nu_0\mathbb{I} = (1-r)m$, that is, $r = 1 - m^{-1} = \rho$. The first part of the lemma is proved.

Now let condition (MD$'$) be satisfied and $\varepsilon < 1$. Then $h \in \mathfrak{N}$ and $RD \in \mathcal{B}$. As above from (7.18) we obtain the dual to (7.19) equation

$$(I - Q)h = (rI - Q)\mathbb{I} \tag{7.21}$$

with the constant $r = 1 - \pi D(\mathbb{I} - h)$. Equation (7.21) implies for all t

$$h = 1 - (1-r)\left(I + \cdots + Q^{t-1}\right)\mathbb{I} - Q^t(1-h). \tag{7.22}$$

Since, by assumption, $h \in \mathfrak{N}$ and $\pi \in \mathfrak{M}$, $\pi h < \infty$. Moreover, from (M2) and Lemma 7.1 it follows that the integrals $\pi Q^t h$ are uniformly bounded. Thus, the monotone sequence of functions $h_t = \sum_{s<t} Q^s\mathbb{I}$ converges in average with respect to the measure π to the function $h_0 = \sum_{s\geq 0} Q^s\mathbb{I}$. Here $\pi h_0 = \sum \pi Q^s\mathbb{I} = \mathsf{E}_\pi\tau = m < \infty$.

Then (7.22) implies that the functions $Q^t(\mathbb{I} - h)$ are monotonically convergent as $t \to \infty$ to the limit $\overline{h} = \mathbb{I} - h - (1-r)h_0$. So, $\overline{h}$ is invariant for Q and from $|\overline{h}| \leq Q|\overline{h}| \leq \mathsf{P}|\overline{h}|$ and $\pi|\overline{h}| = \pi\mathsf{P}|\overline{h}|$ we find that π-almost everywhere $|\overline{h}| = Q|\overline{h}|$. Then by Lemma 7.1

$$\pi DQ^t|\overline{h}| = \pi Q^t|\overline{h}| - \pi Q^{t+1}|\overline{h}| = \pi Q^t\left(|\overline{h}| - Q|\overline{h}|\right) = 0$$

for all $t \geq 0$. The sequence of measures

$$\sum_{s<t} \pi DQ^s = \sum_{s<t} \pi(P-Q)Q^s = \sum_{s<t} \pi\left(Q^s - Q^{s+1}\right) = \pi - \pi Q^t$$

converges monotonically to π, since $\sum \pi Q^t(E) = \pi h_0 < \infty$ and $\pi Q^t(E) \to 0$. Therefore, $\pi|\overline{h}| = \lim \sum_{s<t} \pi DQ^s|\overline{h}|$ and π-almost everywhere $\overline{h} = 0$. Then $\pi\overline{h} = 0$, too. Thus, $1 - m(1-r) = \pi(\mathbb{I} - h - (1-r)h_0) = \pi\overline{h} = 0$ by the definition of $\overline{h}$. The equality $\pi h = 0$ immediately follows from (7.18) and from the relation $\pi R = 0$. So, $r = 1 - m^{-1}$ and the lemma is proved. □

For any distribution α we introduce additionally to (7.15), (7.16) the following indices

$$\varepsilon_{q\alpha} = \sup_{t\geq 0} \left\|\alpha R Q^t \Delta'\right\| \cdot \|\Delta \mathbb{I}\|, \qquad q_\alpha = \sup_{t\geq 0} \left\|\alpha Q^t\right\|,$$
$$\varepsilon'_{q\alpha} = \sup_{t\geq 0} \left\|\alpha Q^t R \Delta'\right\| \cdot \|\Delta \mathbb{I}\| \leq \varepsilon'_0 q_\alpha. \tag{7.23}$$

The index $\varepsilon_{q\alpha}$ is defined for $\alpha \in \mathfrak{M}_0$ and the index $\varepsilon'_{q\alpha}$ is well defined under (MD′).

THEOREM 7.3. *Let either* (MD) *and* $\alpha \in \mathfrak{M}_0$ *or* (MD′) *and* $\alpha \in \mathfrak{M}$ *be satisfied. Assume that* $\varepsilon < 1$. *Then* $m = \mathsf{E}_\pi \tau < \infty$ *and*

$$\sup_{t\geq 0} \left|\mathsf{P}_\alpha(\tau > t) - \rho^t\right| \leq \frac{2}{1-\varepsilon} \min\left\{\kappa\varepsilon_0 + (1-\varepsilon)\varepsilon_{q\alpha}, \varepsilon'_{q\alpha}\right\}. \tag{7.24}$$

Moreover, for $\alpha = \pi$, $\varepsilon_{q\pi} = 0$ *and* $\varepsilon'_{q\pi} \leq \varepsilon'_0 \varkappa$, *that is, the numerator in* (7.24) *does not exceed*

$$2 \min\{\kappa\varepsilon_0, \varkappa\varepsilon'_0\}.$$

If conditions (MD) *and* (MD′) *are fulfilled simultaneously and* $q < \infty$ *then*

$$\sup_{t\geq 0} \left|\mathsf{P}_\alpha(\tau > t) - \rho^t\right| \leq 4 \frac{\varepsilon_0 \varepsilon'_0}{(1-\varepsilon)^2} \kappa q \varkappa. \tag{7.25}$$

REMARK 7.1. Along with (7.24), a slightly precise estimate is valid. This estimate is obtained from (7.24) by replacing of the fractions $\kappa\varepsilon_0/(1-\varepsilon)$ and $\varepsilon'_{q\alpha}/(1-\varepsilon)$ by $\kappa\|\nu\|$ and $\varepsilon'_\alpha = \sup |\alpha Q^t h|$ respectively, where the measure ν and the function h satisfy (7.6), (7.7) and are calculated by (7.18). Besides, the factor 2 in (7.24) may be omitted if the sequence $\alpha Q^t h$ is nonnegative.

REMARK 7.2. A simple computation shows that, under assumptions of Theorem 7.3,

$$\sup_{t\geq 0} \left|\rho^t - \exp\{-(1-\rho)t\}\right| \leq 1 - \rho = \frac{1}{m},$$

where

$$\frac{1}{m} = O\left(\|\Delta \Pi \Delta'\|\right)$$

as $D \to 0$, since from (7.17) follows that

$$\left|\frac{1}{m} - p\right| \leq \frac{\varepsilon}{1-\varepsilon} \|\Delta \Pi \Delta'\| \tag{7.26}$$

with $p = \pi D \mathbb{I} \leq \|\Delta \Pi \Delta'\|$.

In particular, inequality (7.26) implies that m^{-1} in the definition $\rho = 1 - m^{-1}$ is equivalent to p, so that $mp \to 1$ in the scheme of series provided that the chain X is uniformly ergodic and $D \to 0$, $m \to \infty$ in such a way that either $m = o\left(\|D\|^{-2}\right)$ or $\|D\|^2 = o(p)$.

REMARK 7.3. Under the assumption that $\Delta' = I$ or $\Delta = I$ the index ε_0 in (7.24) or, respectively, ε'_0 in (7.25) may be replaced by $\|\pi D R\|$ or $\|R D \mathbb{I}\|$, respectively.

COROLLARY 7.2. *Let condition* (MD) *hold true, and* $\alpha \in \mathfrak{M}_0$. *Then the following asymptotic expansion is valid*

$$\sup_{s\geq 0} \left|\mathsf{P}_\alpha(\tau/m > s) - \exp\{-s\}\right| = O\left(\|D\Pi\| + \|DR\|\right) \tag{7.27}$$

provided that the right-hand side tends to zero. The expansion (7.27) *is uniform with respect to* $\alpha \in \mathfrak{M}_0$ *with bounded* $\|\alpha R\|$, *to all basic chains with the bounded norm* $\|\Pi\|$, *and to all perturbations* D *with* $\sup q < \infty$.

PROOF OF COROLLARY 7.2. Letting in (7.14) $\Delta' = I$ and using (7.24) we arrive at $\kappa\varepsilon_0 + \varepsilon_{q\alpha} = O\left(\|D\Pi\| + \|DR\|\right)$ according to (7.15) and (7.23), since $\|D\mathbb{I}\| = \|D|\pi|\|/\|\pi\|$ and $\|\pi\|$ is bounded by condition. Furthermore, it follows from the convergence of (7.27) to zero that the denominator in (7.24) is isolated from zero. Moreover, one can use $\exp\{-tm\}$ in (7.24) instead of ρ^t to within $O(m^{-1}) = O(\|D\Pi\|)$ as was showed in Remark 7.2. Finally, substituting $t = sm$ into (7.24) we obtain (7.27). The corollary is proved. □

COROLLARY 7.3. *Let the basic chain* X *be uniformly ergodic in the norm* $\|\cdot\|$ *and not vary in the scheme of series. Then*

$$\sup_{t\geq 0} \left|\mathsf{P}_\pi(\tau > t) - \rho^t\right| = O\left(\|D\|^2\right) \tag{7.28}$$

as $\|D\| \to 0$ *provided that* $q < \infty$.

PROOF OF COROLLARY 7.3. By Theorem 1.3, $\mathfrak{M}_0 = \mathfrak{M}$ and $\mathfrak{N}_0 = \mathfrak{N}$, since the potential R is bounded. Then conditions (MD) and (MD′) are satisfied for all $D \in \mathcal{B}$ and the indices $\varepsilon_0 + \varepsilon_0' = O(\|D\|)$ as $\|D\| \to 0$. Substituting these relations into (7.25) we arrive at (7.28). The corollary is proved. □

Note also that a simple example of the chain X with two states, that is $E = \{0, 1\}$, $p_{ij} = 1 - \delta_{ij}$, $D_{ij} = \varepsilon\delta_{i0}\delta_{j1}$, shows that the approximation order in (7.27) and (7.28) with respect to D is right and equals to $O(\varepsilon)$ for $\alpha_j = \delta_{j0}$ and to $O(\varepsilon^2)$ for $\alpha_j = \pi_j = 1/2$.

PROOF OF THEOREM 7.3. According to Lemma 7.2 and Theorem 7.1 representation (7.8) provided that (MD) is satisfied and representation (7.9) holds true if (MD′) holds true with the measure ν and the function h given by (7.18). It is easy to check that under assumptions of the theorem

$$\|\nu\| \leq \frac{\varepsilon_0}{1-\varepsilon}, \qquad \|h\| \leq \frac{\varepsilon_0'}{1-\varepsilon}. \tag{7.29}$$

Moreover, if $\Delta' = I$ then one may replace ε_0 in (7.29) by $\|\pi DR\|$, since the operators $\Delta R = DR$ and $(I + \Delta R\Delta') = I + DR$ commute and

$$\|\nu\| = \left\|\pi DR(I + DR)^{-1}\right\| \leq \frac{\|\pi DR\|}{1-\varepsilon}.$$

Analogously, if $\Delta = I$ then one may replace ε_0' in (7.29) by $\|RD\mathbb{I}\|$, since

$$\|h\| = \left\|(I + RD)^{-1}\, RD\mathbb{I}\right\| \leq \frac{\|RD\mathbb{I}\|}{1-\varepsilon}.$$

Since ε_0 and ε_0' are involved into expressions above only by means of (7.29), Remark 7.3 is valid.

Assume that condition (MD′) is satisfied. Multiplying (7.9) by the measure α and applying (6.35), we estimate the left-hand side of (7.24) as follows

$$\begin{aligned}\left|\alpha[Q]\mathbb{I} - G\right|_0 &= \left| - G(\alpha h) + (\delta - g) * \alpha[Q]h\right|_0 \le \left|\alpha[Q]h\right|_0 + \left|G(\alpha h) + g[Q]h\right|_0 \\ &\le 2\left|\alpha[Q]h\right|_0 = 2\varepsilon_\alpha' = 2\sup_{t\ge 0}\left|\alpha Q^t R\Delta'(I + \Delta R\Delta')^{-1}\Delta\mathbb{I}\right| \\ &\le \frac{2}{1-\varepsilon}\sup_{t\ge 0}\left\|\alpha Q^t R\Delta'\right\| \cdot \|\Delta\mathbb{I}\| = 2\frac{\varepsilon_{q\alpha}'}{1-\varepsilon}.\end{aligned} \tag{7.30}$$

This proves the second part of estimate (7.24).

Note that if $\alpha[Q]h$ on the right hand side of the first inequality in (7.30) is nonnegative then it may be replaced by the maximum of summands so that the factor 2 in (7.30) may be omitted.

Let both conditions (MD) and (MD′) be satisfied. Then representations (7.8) and (7.9) are valid. Multiplying (7.9) by the measure $\pi \in \mathfrak{M}$ and obtaining the sequence $\pi[Q]h$ from representation (7.8) multiplied by the function $h \in \mathfrak{N}$, we get

$$\pi[Q]\mathbb{I} - G = -(\nu h)G * (\delta - g) + (\delta - g) * \nu[Q]h * (\delta - g)$$

where the equality $\pi h = 0$, followed from (7.18), is also used. Then the left-hand side of (7.25) is equal to

$$\begin{aligned}|\pi[Q]\mathbb{I} - G|_0 &\le |\delta - g|_1 \cdot \left| - (\nu h)G + \nu[Q]h * (\delta - g)\right|_0 \\ &\le 2\big(|\nu[Q]h|_0 + |G(\nu h) + g * \nu[Q]h|_0\big) \\ &\le 4|\nu[Q]h|_0 \le 4\|\nu\|\sup_{t\ge 0}\left\|Q^t\right\| \cdot \|h\|.\end{aligned}$$

Substituting (7.29) into the last estimate we arrive at (7.25).

To prove the first part of estimate (7.24) under condition (MD) and for $\alpha \in \mathfrak{M}_0$, we use the following result.

LEMMA 7.3. *Let $\alpha \in \mathfrak{M}_0$ or, respectively, $f \in \mathfrak{N}_0$. Then*

$$\alpha(I - \Pi)[Q]\mathbb{I} = -\alpha RD[Q]\mathbb{I} + \alpha R[Q]D\mathbb{I}, \tag{7.31}$$

$$\pi[Q](I - \Pi)f = \pi D[Q]Rf - \pi[Q]DRf. \tag{7.32}$$

PROOF OF LEMMA 7.3. Observe that $\alpha R \in \mathfrak{M}$ for $\alpha \in \mathfrak{M}_0$. By Definition 1.3 of the potential R we have $\alpha(I - \Pi) = \alpha R(I - P)$. Thus,

$$\alpha(I - \Pi) = \alpha R(I - Q - D) = -\alpha RD + \alpha R(I - Q).$$

Multiplying this equality by the function $Q^t\mathbb{I}$ and noting that $(I-Q)Q^t\mathbb{I} = Q^t(I-Q)\mathbb{I} = Q^t(P\mathbb{I} - Q\mathbb{I}) = Q^tD\mathbb{I}$ we get

$$\begin{aligned}\big(\alpha(I - \Pi)[Q]\mathbb{I}\big)_t &= \alpha(I - \Pi)Q^t\mathbb{I} = -\alpha RDQ^t\mathbb{I} + \alpha R(I - Q)Q^t\mathbb{I} \\ &= -\alpha RDQ^t\mathbb{I} + \alpha RQ^tD\mathbb{I} = (-\alpha RD[Q]\mathbb{I} + \alpha R[Q]D\mathbb{I})_t\,.\end{aligned}$$

Relation (7.32) is derived similarly. The lemma is proved. □

Now we come back to the proof of the theorem. Under conditions (MD) and $\varepsilon < 1$ representation (7.8) is valid by Lemma 7.2. Analogously, under $\alpha \in \mathfrak{M}_0$ representation (7.31) holds true. Then the sequence on the left-hand side of (7.24) is equal to

$$\begin{aligned}\alpha[Q]\mathbb{I} - G &= (\pi[Q] - G\pi)\,\mathbb{I} + \alpha(I - \Pi)[Q]\mathbb{I} \\ &= (\delta - g) * \nu[Q]\mathbb{I} + \alpha R[Q]D\mathbb{I} - \alpha RD[Q]\mathbb{I}.\end{aligned} \tag{7.33}$$

To get (7.33) we used the properties that $\alpha\Pi = \pi$ and $\nu\mathbb{I} = 0$ that follow from definitions of $\Pi = \mathbb{I} \circ \pi$ and ν.

To estimate the right-hand side of (7.33) we note that by Lemma 7.1 the sequences $Q^t\mathbb{I}$ and $\Delta Q^t\mathbb{I}$ do not increase. Since (MD) implies (MD'), $\|Q^t\mathbb{I}\| \le \|\mathbb{I}\| = \kappa$ and $\|\Delta Q^t\mathbb{I}\| \le \|\Delta\mathbb{I}\|$. Therefore,

$$\begin{aligned}|\alpha[Q]\mathbb{I} - G|_0 &\le |\delta - g|_1\|\nu\| \cdot |[Q]\mathbb{I}|_0 + |\alpha R[Q]\Delta'|_0 \cdot \|\Delta\mathbb{I}\| + \|\alpha R\Delta'\| \cdot |\Delta[Q]\mathbb{I}|_0 \\ &\le 2\kappa\|\nu\| + \varepsilon_{q\alpha} + \|\alpha R\Delta'\| \cdot \|\Delta\mathbb{I}\| \le 2\kappa\|\nu\| + 2\varepsilon_{q\alpha}.\end{aligned} \tag{7.34}$$

Substituting inequality (7.29) into (7.34) we obtain the first part of the estimate (7.24). Theorem 7.3 is proved. The assertion of Remark 7.1 evidently arises from (7.30) and (7.34). □

7.3. Approximations of the joint distribution of a Markov moment and a chain

The convergence of the joint distribution of τ and X is essentially connected to asymptotics of powers P^t as $t \to \infty$. For this reason, estimates of such a convergence cannot be expressed only in terms of averaged indices ε, ε_0, ε_0' that are constructed by the potential. In this connection, we introduce some additional indices

$$\varepsilon_{1\alpha} = \sum_{t\ge0} \left\|\alpha\left(P^t - \Pi\right)D\right\|, \qquad \varepsilon_1' = \sum_{t\ge0} \left\|D\left(P^t - \Pi\right)\right\|. \tag{7.35}$$

Recall that methods leading to estimations of indices (7.35) for general Markov ergodic chains are described in Chapter 5.

THEOREM 7.4. *Let either condition* (MD) *be satisfied,* $q < \infty$, $\alpha \in \mathfrak{M}_0$ *or condition* (MD') *be satisfied,* $q_\alpha < \infty$, $\alpha \in \mathfrak{M}$ *and* $\varepsilon < 1$. *Then, for all* $f \in \mathfrak{N}$ *and* $t \ge 0$,

$$\begin{aligned}&\left|\mathsf{E}_\alpha\left(f(X_t), \tau > t\right) - \rho^t(\pi f)\right| \\ &\quad\le \left|\alpha\left(P^t - \Pi\right)f\right| + \frac{\|f\|}{1-\varepsilon}\min\left\{(2\varepsilon_0 + \varepsilon_{1\alpha}(1-\varepsilon))q, (2\varkappa\varepsilon_0' + \varepsilon_1')q_\alpha\right\},\end{aligned} \tag{7.36}$$

where the first summand and $\varepsilon_{1\alpha}$ *are equal to zero for* $\alpha = \pi$.

REMARK 7.4. The domain of the time variable $t \ge 0$ in (7.23), (7.24), (7.35), and (7.36) can be simultaneously replaced by the set $0 \le t \le t_0$ so that the common power of kernels P, Q, and D involved in these relations does not exceed t_0.

Note also that the first summand $|\alpha\left(P^t - \Pi\right)f|$ in (7.36) cannot be omitted, since, for small t, the left-hand side of (7.36) converges as $D \to 0$ to this term. Thus one can expect that the approximation error in (7.36) is small only in the case where both $t \to \infty$ and $D \to 0$.

COROLLARY 7.4. *Let a chain X satisfy condition (C) and be aperiodic and uniformly ergodic in the norm $\|\cdot\|$ and let it be unvarying in the scheme of series. Then, for all $\alpha \in \mathfrak{M}$, $f \in \mathfrak{N}$, the following asymptotic expansion is valid*

$$\left|\mathsf{E}_\alpha\left(f(X_{sm}), \tau > ms\right) - e^{-s}\pi f\right| = O(\|D\|)$$

as $\|D\| \to 0$ and

$$\frac{1}{s} = O\left(\frac{1}{\|D\| \ln \frac{1}{\|D\|}}\right).$$

This expansion is uniform in $\|\alpha\| \leq 1$ and $\|f\| \leq b$ in any domain $s \geq s_0$ for all fixed $s_0 > 0$.

PROOF. Corollary 2.1 implies that the sum $\sum \|P^t - \Pi\|$ is finite and $\|P^t - \Pi\| = O\left(\beta^t\right)$ for some $\beta < 1$. Then the indices ε, ε_0, ε_0', and ε_1' (7.36) are $O(\|D\|)$ for $\Delta = I$ and $Q_\alpha = \sup \|\alpha Q^t\| \leq \sup \|\alpha P^t\| \leq \|\alpha\| \sup \|P^t\|$ is uniformly bounded in α. Thus, the second term on the right-hand side is $O(\|D\|)$. It is easily seen that, for $t = ms$,

$$\left|\alpha\left(P^t - \Pi\right) f\right| \leq ab\left\|P^t - \Pi\right\| = O\left(\beta^{sm}\right) = O(\|D\|)$$

provided

$$\frac{1}{s} = O\left(\frac{m}{\ln \frac{1}{\|D\|}}\right).$$

But the estimate (7.26) shows that $m^{-1} = O(\|D\|)$, that is,

$$\frac{1}{\|D\| \ln \frac{1}{\|D\|}} = O\left(\frac{m}{\ln \frac{1}{\|D\|}}\right).$$

Then the assumptions on the variation of s imply the representation needed. The difference $\rho^t - e^{-s}$ is equal to $O\left(m^{-1}\right) = O(\|D\|)$ by Remark 7.2. The corollary is proved. □

PROOF OF THEOREM 7.4. Let condition (MD) be satisfied and $\varepsilon < 1$. Then, according to the representation (7.8) and the equality of Lemma 6.4, the sequence formed by differences of the left-hand side of (7.36) after subtracting of items $\alpha\left(P^t - \Pi\right) f$ is equal to

$$\begin{aligned}
&\alpha[Q]f - G(\pi f) - (\alpha[P]f - \alpha[\Pi]f) \\
&\quad = \alpha(I - \Pi)[Q]f + \alpha\Pi[Q]f - G(\pi f) - \alpha(I - \Pi)[P]f \\
&\quad = (\pi[Q] - \pi G)f + \alpha(I - \Pi)([Q] - [P])f \\
&\quad = -G(\nu f) + (\delta - g) * \nu[Q]f + \alpha([P] - [\Pi])D * \sigma * [Q]f.
\end{aligned} \tag{7.37}$$

Therefore, according to (6.35) and (6.36) we get

$$\begin{aligned}
&\left|\alpha[Q]f - G(\pi f) - (\alpha[P]f - \alpha[\Pi]f)\right|_0 \\
&\quad \leq |\nu[Q]f|_0 + |G(\nu f) + g * \nu[Q]f|_0 + |\alpha([P] - [\Pi])D|_1 |\nu[Q]f|_0 \\
&\quad \leq (2q\|\nu\| + q\varepsilon_{1\alpha})\|f\|.
\end{aligned}$$

Substituting inequality (7.29) into the last one we obtain the first part of (7.36).

Now let condition (MD′) be satisfied and $\varepsilon < 1$. Transform the left-hand side of (7.37) to the form

$$\begin{aligned}\alpha[Q]f - G(\pi f) - (\alpha[P]f - \alpha[\pi]f) \\ = \alpha[Q]\Pi f + \alpha[Q](I-\Pi)f - G(\pi f) - \alpha[P](I-\Pi)f \\ = (\alpha[Q]\mathbb{1} - G)(\pi f) + \alpha([Q]-[P])(I-\Pi)f \\ = (\alpha[Q]\mathbb{1} - G)(\pi f) + \alpha[Q] * \sigma * D([P]-[\Pi])f,\end{aligned} \tag{7.38}$$

where identity (6.30) is also used.

Estimate, by Theorem 7.3, the first summand on the right-hand side of (7.38). The norm $|\cdot|_0$ of the second summand may be estimated by $|\alpha[Q]|_0 D([P]-[\Pi])f|_1 \le q_\alpha \varepsilon_1' \|f\|$. Then we obtain the second part of (7.36). The theorem is proved. □

The assertion of Remark 7.4 may be established in the same way as that of Remark 6.5. Namely, inequalities (6.35) are (6.36) used in those proofs remain true if the variable t in the definitions of the norms $|\cdot|_i$ runs only over $0 \le t \le t_0$ as was shown above.

COROLLARY 7.5. *Let the moment τ coincide with the first hitting time $\tau_H = \inf(t \ge 1: X_t \in H)$ of a set $H \in \mathcal{E}$. Assume that $\pi(H) \to 0$ in a scheme of series where π is a unique invariant stochastic measure of the chain X.*

Denote by $\overline{R} = RP$ the product of the potential R of the chain with its transition kernel P. This operator, by the definition, is a solution of the system

$$\begin{aligned}\overline{R}(x,A) &= \int_E P(x,dy)\,\overline{R}(y,A) + P(x,A) - \pi(A), \\ \overline{R}(x,E) &= 0,\end{aligned}$$

for $x \in E$, $A \in \mathcal{E}$ and coincides with the Cesaro sum

$$\overline{R}(x,A) = \sum_{t\ge 1}{}^{(c)} \left(P^t(x,A) - \pi(A)\right).$$

Assume further that the kernel $\overline{R}$ has the variation on H that is integrable with respect to the measure π and the variation of the restriction of $\overline{R}$ on H is less than unity:

$$\begin{aligned}r_{\pi H} &= \int_E \pi(dx)\,|\overline{R}|(x,H) < \infty \\ r_{HH} &= \sup_{x\in H} |\overline{R}|(x,H) < 1.\end{aligned} \tag{7.39}$$

Let an initial distribution α of the chain X have a bounded density with respect to π so that $\alpha(A) \le a\pi(A)$ for all $A \in \mathcal{E}$. Then the following inequality is valid

$$\sup_{t\ge 0}\left|\mathsf{P}_\alpha(\tau_H > t) - \left(1 - \frac{1}{m_H}\right)^t\right| \le 2a\frac{r_{\pi H}}{1 - r_{HH}}, \tag{7.40}$$

where the mean value $m_H = \mathsf{E}_\pi \tau_H$ *is equal to*

$$m_H = \left(\sum_{n \geq 0} (-1)^n \int_H \pi(dx) \int_H \overline{R}(x_1, dx) \cdots \int_H \overline{R}(x_n, H) \right)^{-1} \tag{7.41}$$

and satisfies the estimate

$$\left| \frac{1}{m_H} - \pi(H) \right| \leq \pi(H) \frac{r_{HH}}{1 - r_{HH}}.$$

If, instead of the finiteness of $r_{\pi H}$, *we suppose that the potential* R *has a bounded variation for* $x \in H$

$$r_H = 1 + \sup_{x \in H} |\overline{R}|(x, E) < \infty \tag{7.42}$$

and the distribution α *additionally satisfies the inequality*

$$|\mathsf{P}_\alpha(X_t \in A) - \pi(A)| \leq b(1 - r_\alpha)^t$$

for $A \in \mathcal{E}$, $t \geq 0$, *and some* $r_\alpha > 0$ *then, for all* $t > t_0 = r_\alpha^{-1} \ln^+ (b/a\pi(H))$,

$$\begin{aligned} &\left| \mathsf{P}_\alpha(\tau_H > t, X_t \in A) - \pi(A) \left(1 - \frac{1}{m_H} \right)^t \right| \\ &\qquad \leq \pi(H) \left[2 \frac{r_H}{1 - r_{HH}} + \frac{a}{r_\alpha} \ln \left(1 + b \frac{e}{a\pi(H)} \right) \right] \end{aligned} \tag{7.43}$$

uniformly with respect to $A \in \mathcal{E}$. *Here, for* $\alpha = \pi$, t_0 *and the second summand in* (7.43) *are equal to zero, since* $b = 0$.

REMARK 7.5. Conditions (7.39) for the potential R may be fulfilled even if the chain X is not uniformly mixing. In order that $r_{\pi H}$ be convergent to zero as $\pi(H) \to 0$, it is sufficient, for example, that the series $\sum (P^t(x, A) - \pi(A))$ be convergent in $L_1(\pi)$ uniformly with respect to $A \in H \cap \mathcal{E}$ and uniformly in the scheme of series, since, for all $n \geq 1$,

$$r_{\pi H} \leq 2n\pi(H) + \int_E \pi(dx) \sup_{A \in H \cap \mathcal{E}} \left| \sum_{t \geq n} (P^t(x, A) - \pi(A)) \right|.$$

PROOF. Let $\tau = \tau_H$. Then the corresponding kernel $D = P - Q$ is equal to

$$D(x, A) = \mathsf{P}_x(\tau_H = 1, X_1 \in A) = \mathsf{P}_x(X_1 \in A \cap H) = \mathsf{P}I_H(x, A),$$

where $I_H(x, A) = \mathbb{I}_{\{x \in A \cap H\}}$. Since $I_H = I_H^2$, $D = PI_H = PI_H \cdot I_H$ and elements of the factorization in (7.14) may be chosen as $\Delta' = PI_H$, $\Delta = I_H$.

To derive (7.40) we take, as $\mathfrak{M}$, the space of all measures on $\mathcal{E}$ having absolutely continuous restrictions on $E \setminus H$ with respect to π and finite norms of the form

$$\|\mu\| = \max \left(\sup_{x \in E \setminus H} \frac{d\mu}{d\pi}(x), \sup_{A \in H \cap \mathcal{E}} |\mu(A)| \right). \tag{7.44}$$

Since $r_{\pi H}$ is finite, the operator $R\Delta' = RPI_H = \overline{R}I_H$ is bounded on $\mathfrak{M}$ because $\mu \in \mathfrak{M}$ and $\|\mu\| \leq 1$ imply $|\mu|(A) \leq \pi(A)$ and $|\mu|(H) \leq 2$ for all $A \in \mathcal{E} \cap (E \setminus H)$. So,

$$\begin{aligned}\|\mu R I_H\| &= \sup_{A \in \mathcal{E} \cap H} |\mu \overline{R}(A)| \leq \int_{E \setminus H} |\mu|(dx)|\overline{R}|(x, H) + \int_H |\mu|(dx)|\overline{R}|(x, H) \\ &\leq \int_{E \setminus H} \pi(dx)|\overline{R}|(x, H) + 2 \sup_{x \in H} |\overline{R}|(x, H) \leq r_{\pi H} + 2r_{HH}.\end{aligned}$$

Thus condition (MD′) is satisfied and the second part of estimate (7.24) is valid. Moreover, by the conditions

$$\begin{aligned}\varepsilon = \|\Delta R \Delta'\| &= \|I_H \overline{R} I_H\| = \sup \left(|\mu I_H \overline{R}(A)|, A \in H \cap \mathcal{E}, \|\mu I_H\| \leq 1\right) \\ &\leq \sup \left(|\overline{R}|(x, H), x \in H\right) = r_{HH} < 1.\end{aligned} \tag{7.45}$$

The upper bound in (7.45) follows from the estimate

$$\begin{aligned}|\mu Q(A)| &= \left|\left(\mu^+ - \mu^-\right)\left(Q^+ - Q^-\right)(A)\right| \\ &\leq \max\left[\left(\mu^+ Q^+ + \mu^- Q^-\right)(A), \left(\mu^- Q^+ + \mu^+ Q^-\right)(A)\right] \\ &\leq \max\left[\mu^+(H), \mu^-(H)\right] \sup_{x \in H}\left[Q^+(x, H) + Q^-(x, H)\right] \\ &= \sup_{A \in H \cap \mathcal{E}} |\mu(A)| \sup_{x \in H} |Q|(x, H),\end{aligned} \tag{7.46}$$

that is uniform in $A \in H \cap \mathcal{E}$, where $\mu^\pm$ and $Q^\pm$ are nonnegative components in the Hahn decomposition of the measure μ and the kernel Q, respectively.

In view of $\alpha \leq a\pi$ and Lemma 7.1, $\alpha Q^t \leq a\pi Q^t \leq a\pi$. Furthermore, the dual norm on $\mathfrak{N}$ is of the form

$$\|f\| = \pi\left|I^H f\right| + \sup_{x \in H} |f(x)|.$$

Therefore the following estimates are valid

$$\begin{aligned}\varepsilon'_{q\alpha} &= \sup_{t \geq 0} \left\|\alpha Q^t \overline{R} I_H\right\| \cdot \|I_H \mathbb{I}\| = \sup_{t \geq 0} \sup_{A \in H \cap \mathcal{E}} \left|\alpha Q^t \overline{R}(A)\right| \\ &\leq \sup_{t \geq 0} \int_E \alpha Q^t(dx)|\overline{R}|(x, H) \leq a \int_E \pi(dx)|\overline{R}|(x, H) = a r_{\pi H}.\end{aligned}$$

Substituting this estimate together with (7.45) into (7.24) we arrive at inequality (7.40).

Equality (7.41) is immediately derived from (7.17).

Now let $r_{HH} < 1$ and $r_H < \infty$. To obtain (7.43) we take the space of all finite measures on $\mathcal{E}$ with the semivariation norm $\|\mu\| = \sup(|\mu(A), A \in \mathcal{E})$ as $\mathfrak{M}$ and use the first part of estimate (7.36) with functions from the set $\{\mathbb{I}_A(x), A \in \mathcal{E}\}$. Condition (MD) is satisfied because the operator $\Delta R = I_H R$ is bounded. Indeed, inequality (7.46) applied to the set $H = E$ and to the kernel $S = I_H \overline{R}$ implies

$$\|\mu I_H \overline{R}\| \leq \|\mu\| \sup_{x \in E} |I_H \overline{R}|(x, E) = \|\mu\|(r_H - 1).$$

Besides, from the definition of $\overline{R}$ it is easy to obtain that $\overline{R} = R - I + \Pi$. Then

$$\|I_H R\| \le \|I_H \overline{R}\| + \|I - \Pi\| \le r_H - 1 + 1 = r_H$$

and $I_H R \in \mathcal{B}$.

Further, the index $\varepsilon = \|\Delta R \Delta'\| = \|I_H \overline{R} I_H\|$ coincides with that calculated in (7.45), since the kernel $I_H \overline{R} I_H$ is equal to its restriction on $(H, H \cap \mathcal{E})$ and the norm under consideration coincides with the norm in (7.44) on all measures that are concentrated on H. Thus $\varepsilon = r_{HH} < 1$.

Since $Q \le P$ is substochastic and

$$\varepsilon_0 = \|\pi\Delta'\| \cdot \|\Delta R\| = \|\pi I_H\| \cdot \|I_H R\| \le \pi(H) r_H,$$
$$q = \sup \|Q^t\| \le \sup \|P^t\| = 1,$$

the left hand side of (7.43) does not exceed

$$2\pi(H)\frac{r_H}{1 - R_{HH}} + e_t(\alpha) \tag{7.47}$$

for all $t \ge 0$ because of Remark 7.4 with $\varepsilon_{1\alpha}$ and (7.36) , where

$$e_t(\alpha) = \sum_{s<t} \|\alpha (P^s - \Pi) P I_H\| + \left\|\alpha P^t - \pi\right\|.$$

Let t_0 defined in (7.43) be equal to $n + c$ with $n = [t_0] < t$ by the condition and $0 \le c < 1$. Let us estimate summands of the sum $e_t(\alpha)$ as follows

$$\begin{aligned}\|\alpha (P^s - \Pi) P I_H\| &= \left\|\alpha P^{s+1} I_H - \pi I_H\right\| \le \max\left\{\alpha P^{s+1}(H), \pi(H)\right\} \\ &\le \max\left\{a\pi P^{s+1}(H), \pi(H)\right\} = \pi(H) \max\{a, 1\} = a\pi(H)\end{aligned}$$

for $s < n$,

$$\|\alpha (P^s - \Pi) P I_H\| \le \|\alpha P^s - \pi\| \cdot \|P I_H\| \le b(1 - r_\alpha)^s$$

for $s > n$, and finally,

$$\|\alpha (P^n - \Pi) P I_H\| \le ca\pi(H) + (1 - c) b (1 - r_\alpha)^n$$

for $s = n$. Therefore, for all $t > t_0$,

$$\begin{aligned}e_t(\alpha) &\le na\pi(H) + ca\pi(H) + (1 - c)b(1 - r_\alpha)^n + b(1 - r_\alpha)^{n+1} + \cdots \\ &\quad + b(1 - r_\alpha)^{t-1} + b(1 - r_\alpha)^t \\ &= t_0 a\pi(H) + b(1 - r_\alpha)^{t_0 - c}\frac{1 - cr_\alpha}{r_\alpha} - b\frac{1 - r_\alpha^{t+1}}{r_\alpha} \\ &\le t_0 a\pi(H) + b\frac{(1 - r_\alpha)^{t_0}}{r_\alpha} \le t_0 a\pi(H) + \frac{b}{r_\alpha} e^{-t_0 r_\alpha},\end{aligned} \tag{7.48}$$

since $1 - rc \le (1 - r)^c \le \exp\{-rc\}$ for $0 \le c < 1$.

It remains to note that the right hand side of (7.48) is equal to

$$\frac{b}{r_\alpha} \le \pi(H)\frac{a}{r_\alpha}\ln\left(1+\frac{be}{a\pi(H)}\right)$$

for $t_0 = 0$, that is, for $b \le a\pi(H)$, since $u \le \ln(1+eu)$ for $0 \le u \le 1$ and the right-hand side of (7.48) for $t_0 > 0$ coincides, by the definition of t_0, with the expression

$$a\pi(H)\frac{1+t_0 r_\alpha}{r_\alpha} = \pi(H)\frac{a}{r_\alpha}\left(1+\ln\frac{b}{a\pi(H)}\right) \le \pi(H)\frac{a}{r_\alpha}\ln\left(1+\frac{be}{a\pi(H)}\right).$$

So, according to (7.48), for all $t_0 \ge 0$ and $t > t_0$, $e_t(\alpha)$ does not exceed the second summand on the right-hand side of (7.43). Substituting this inequality into estimate (7.47) of the left-hand side of (7.43) we arrive at estimate (7.43). The corollary is proved. □

EXAMPLE 7.1. Let a sequence $Y = (Y_t, t \ge 0)$ consist of random variables that are conditionally independent under a given trajectory of a chain $X = (X_t, t \ge 0)$. The random variables Y_t are supposed to have the conditional distribution

$$\mathsf{P}\left(Y_t = 0/X\right) = \delta g(X_t) = 1 - \mathsf{P}\left(Y_t = 1/X\right).$$

Consider a random moment $\tau_\delta = \inf(t \ge 1 : Y_t = 0)$. This moment can be interpreted as a killing moment for the chain X that is killed, independently of other events, after a jump to a state $x \in E$ with the probability $\delta g(x)$.

Assume that the function g is measurable and bounded, that is, $0 \le g(x) \le 1$, $x \in E$. Let $\delta \to 0$ and the chain X possesses the uniform mixing property

$$\sup_{x,y\in E}\sup_{A\in\mathcal{E}}|P^n(x,A) - P^n(y,A)| \le 1 - r \tag{7.49}$$

for some $n \ge 1$ and $r > 0$.

Then, for all $\delta < r/n$,

$$\left|\mathsf{P}_x(\tau_\delta > t, X_t \in A) - \pi(A)\left(1 - \frac{1}{m_\delta}\right)^t\right| \le 2\delta n\left(\frac{1}{r} + \frac{G}{r - \delta n}\right) \tag{7.50}$$

uniformly with respect to $x \in E$, $A \in \mathcal{E}$, and $t > nr^{-1}\ln(2r/\delta n)$, where

$$m_\delta = \mathsf{E}_\pi\,\tau_\delta, \qquad G = \int_E g(x)\,\pi(dx)$$

and

$$\left|\frac{1}{m_\delta} - \delta G\right| \le 2\delta^2 n^2 \frac{G}{r(r - 2\delta n)}. \tag{7.51}$$

The proof of (7.50) is based on (7.36). The perturbation kernel in the case under consideration is equal to $D(x,A) = \delta g(x)P(x,A)$. Let us take in (7.14) $\Delta = D$,

$\Delta' = I$ and define $\|\mu\| = \sup(|\mu(A)|, A \in \mathcal{E})$. Then $\|D\| \leq \delta$ and from (7.49) we conclude that

$$\|\alpha P^t - \pi\| \leq (1-r)^{[t/n]} \leq n\frac{\delta}{r}$$

for appropriate t and all α. Using Theorem 3.4 we get

$$\varepsilon_{1\alpha} \leq n\frac{\delta}{r},$$
$$\varepsilon_0 \leq \sum \|\pi D\,(P^t - D)\| \leq \pi D(E)\frac{n}{r} = \delta G\frac{n}{r},$$
$$\varepsilon \leq \sum \|D\| \cdot \|P^t - D\| \leq \delta\frac{n}{r}.$$

Substituting these inequalities into the right-hand side of (7.36) we obtain (7.50).

CHAPTER VIII

STABILITY OF WAITING TIMES IN A QUEUING SYSTEM

In this chapter, we consider the stability problem for a Markov chain X related to waiting times in a queuing system $GI/GI/1$ under the assumption that perturbations of the chain are of the same structure. We obtain estimates obtained for stability of finite-time and ergodic distributions that have maximal, and namely, first order of accuracy with respect to perturbations of a control sequence. Corresponding asymptotic expansions are also given.

Consider a Markov chain with values in the space $E = \mathbb{R}_+$ equipped with the Borel σ-field $\mathcal{E} = \mathcal{B}_+ = \mathcal{B}(\mathbb{R}_+)$. Assume that this chain is given for $t \geq 0$ by the recurrent relations

$$X_{t+1} = (X_t + \xi_{t+1})^+ , \qquad (8.1)$$

where $x^+ = \max\{0, x\}$. The control sequence $\{\xi_t, t \geq 1\}$ consists of independent identically distributed random variables with the common distribution $F(B) = \mathsf{P}(\xi_t \in B)$ and with the finite mean $m = \mathsf{E}\,\xi_t$.

It is known, [see Borovkov (1972), Chapter 4], that a chain X is ergodic and has a unique invariant stochastic measure π provided that $m < 0$. In what follows we assume this condition throughout.

The transition kernel P of a chain X is for $x \geq 0$, $B \in \mathcal{B}_+$, of the form

$$P(x, B) = \int_{-\infty}^{+\infty} F(dy)\,(\alpha(B)\mathbb{I}_{y \leq -x} + \mathbb{I}_{0 < x+y \in B}) = h(x)\alpha(B) + T(x, B), \quad (8.2)$$

where α is a distribution concentrated at zero, $h(x) = F(-\infty, -x)$, and the kernel $T(x, B) = \mathsf{P}(0 < \xi_1 + x \in B)$ is nonnegative. Therefore the chain X satisfies the condition (C) for $n = 1$.

Along with X we also consider the chain $Y = (Y_t)$ on $(E, \mathcal{E})$ that is constructed by a control sequence $\{\eta_t, t \geq 1\}$ of independent random variables with a common distribution G in the same way as X is constructed by F.

In what follows, measures μ on $\mathcal{B}_+$ are extended on $\mathcal{B}(\mathbb{R}_-)$ by putting $\mu(B) = 0$ for $B \in \mathcal{B}(\mathbb{R}_-)$.

8.1. Uniform ergodicity and strong stability

For measures on $\mathcal{B}_+$, we introduce the family of norms of the form

$$\|\mu\|_c = \int_0^{\infty} \exp\{cx\}\,|\mu|(dx). \qquad (8.3)$$

These are $\|\cdot\|_v$-norms in (1.5) with functions $v_c(x) = \exp\{cx\}$, $x \geq 0$. Let

$$\mathfrak{M}_c = \{\mu \in m\mathcal{B}_+ : \|\mu\|_c < \infty\}.$$

THEOREM 8.1. *Let the Cramér condition*

$$\mathsf{E}\exp\{c\xi_1\} < \infty \tag{8.4}$$

is satisfied for some $c > 0$. The chain X is aperiodic and uniformly ergodic in a weighted variation norm $\|\cdot\|_c$ if and only if

$$\rho(c) = \mathsf{E}\exp\{c\xi_1\} < 1. \tag{8.5}$$

The corresponding set is nonempty and coincides with an interval $(0, d)$ because of (8.4) and condition $m < 0$.

PROOF. Condition (8.4) implies that the operator P is bounded on $\mathfrak{M}_c$. According to (8.2) and Theorem 2.3, the assertion of the theorem is equivalent to the relation

$$\begin{aligned} 1 > \left\|T^k\right\|_c &= \sup_{x\geq 0}\exp\{-cx\}T^k v_c(x) \\ &= \sup_{x\geq 0}\exp\{-cx\}\,\mathsf{E}\left(\exp\{cS_k + cx\}, S_1 > -x, \dots, S_k > -x\right) \\ &= \mathsf{E}\exp\{cS_k\} = (\rho(c))^k \end{aligned} \tag{8.6}$$

for some $k \geq 1$, where $S_k = \xi_1 + \cdots + \xi_k$. It remains to observe that the function $\rho(c)$ is convex, $\rho(0) = 1$, and $\rho'(0) = m < 0$. Thus $d = \sup\{c \geq 0 \colon \rho(c) < 1\}$ is positive. The theorem is therefore proved. □

THEOREM 8.2. *If condition (8.4) is satisfied for any c with (8.5) then the following estimate is valid*

$$\|\pi\|_c = \int_0^\infty \exp\{cx\}\,\pi(dx) \leq \frac{1}{1-\rho(c)} \tag{8.7}$$

and, as $t \to \infty$, the asymptotic expansion

$$\mathsf{E}_x(f(X_t), X_t \in B) - \int_B f(x)\,\pi(dx) = O\left(\exp\{cx\}t\theta_0^t\right) \tag{8.8}$$

holds true uniformly with respect to $x \geq 0$, $B \in \mathcal{B}_+$, and all functions f such that $|f(x)| \leq \exp\{cx\}$. Here the constant $\theta_0 < 1$ is defined by (3.29), (3.31), (3.32) where $\rho(c)$ is used instead of ρ.

The proof of the theorem immediately follows from Corollary 3.2, estimate (8.6) for $k = 1$, and from estimate (3.30) of Theorem 3.6. □

To obtain asymptotic expansions in stability theorems for a chain X, denote by

$$\delta(B) = G(B) - F(B)$$

the perturbation of a distribution F. Then the corresponding kernel $D = Q - P$ is determined by (8.2) by using δ instead of F:

$$D(x, B) = \int_{-\infty}^\infty \delta(dy)\left(\alpha(B)\mathbb{1}_{y\leq -x} + \mathbb{1}_{0<x+y\in B}\right). \tag{8.9}$$

Using (1.5), it is easy to check that $\|D\|_c \le |\delta|_c$, where

$$\begin{aligned}|\delta|_c &= \sup_{x\ge 0}\left(|\delta(-\infty,-x)|\exp\{-cx\} + \int_x^\infty \exp\{cy\}|\delta|(dy)\right)\\ &\le \int_{-\infty}^\infty \exp\{cy^+\}\,|\delta|(dy).\end{aligned} \tag{8.10}$$

For $x \ge 0$, $B \in \mathcal{B}_+$, define the kernel

$$R_0(x,B) = \sum_{t\ge 0} \mathsf{P}(\tau_x > t, S_t + x \in B), \tag{8.11}$$

where $S_t = \xi_1 + \dots + \xi_t$, $\tau_x = \inf\{t \ge 1 \colon S_t + x \le 0\}$. It follows from the definition of $T(x,B)$ that

$$R_0 = \sum_{t\ge 0} T^t = (I-T)^{-1}.$$

LEMMA 8.1. *Let $m < 0$. Then the function $\varphi(x) = R_0(x,E)$ is bounded for $x \ge 0$ and the generalized potential R of the chain X coincides with the σ-finite kernel*

$$R = (I-\Pi)R_0(I-\Pi), \tag{8.12}$$

where $\Pi = \mathbb{I}\circ\pi$ is the stationary projector of the chain X.

The proof of the lemma immediately follows from Theorem 5.5 applied to the set $C = \{0\}$ for $n = 1$. □

It is clear that $R_c = R_0$ under notations (5.12). Thus condition (B) follows from the boundedness of the function φ that is established below in Lemma 8.2. The σ-boundedness of the kernel R_0 is derived in Theorem 5.5. Representation (8.12) follows from (5.24) for $n = 1$ and from equality (5.29), where the kernel R_q is equal to zero, since it coincides with the potential of the Markov chain at the single-point set $C = \{0\}$.

Identities (8.9) and (8.12) imply, in particular, that the kernel $\Delta = DR$ involved in (5.6) is equal to

$$\Delta(x,B) = \int_{-x}^\infty \delta(dy)\left(R_0(x+y,B) - \pi(B)T_0(x+y,E)\right), \tag{8.13}$$

since $D\Pi = 0$. Furthermore, it follows from Lemma 3.1 that

$$\alpha R_0(B) = \alpha(I-T)^{-1}(B) = \pi(B)/\pi h,$$

whence

$$\pi(B) = (R_0(0,E))^{-1} R_0(0,B) = \frac{1}{q}\sum_{t\ge 0}\mathsf{P}(\tau_0 > t, S_t \in B), \tag{8.14}$$

where $q = R_0(0,E) = \mathsf{E}\,\tau_0 = 1/\pi(\{0\})$.

By Theorem 8.1, the potential R is bounded on $\mathfrak{M}_c$ and $\|D\|_c \le |\delta|_c$ as we showed above. Thus, $\|DR\|_c = \|\Delta\|_c = O(|\delta|_c)$ as $\delta \to 0$. Therefore, Theorems 4.3 and 4.4 give us the following result.

THEOREM 8.3. *Let (8.5) be satisfied with a constant c. Then for any perturbation $\delta = G - F$ with sufficiently small norm $|\delta|_c$ given by (8.10), the chain Y has a unique invariant stochastic measure $\nu \in \mathfrak{M}_c$ and for each $n \geq 1$, as $|\delta|_c \to 0$, the following asymptotic expansions are valid*

$$\int_0^\infty f(x)\left(\nu - \pi - \pi\Delta - \cdots - \pi\Delta^{n-1}\right)(dx) = O\left(|\delta|_c^n\right), \tag{8.15}$$

$$\mathsf{E}_x f(Y_t) - \mathsf{E}_x f(Y_t) = O(|\delta|_c \exp\{cx\}) \tag{8.16}$$

uniformly in $x \geq 0$, $t \geq 0$, all measurable functions f with $|f(x) \leq \exp\{cx\}$, and all chains X with indices $\rho(c)$ isolated from the unity.

Note that Theorems 3.6 and 3.8 also give explicit estimates for the left-hand sides of (8.15) and (8.16).

8.2. Stability under relatively bounded perturbations

Theorem 8.1 shows us that the chain X is not uniformly ergodic in the total variation norm $\|\mu\| = |\mu|(E)$. For this reason one cannot expect that (8.14) and (8.15) are valid for $c = 0$. But in this case the results of Theorems 5.1–5.5 are applicable.

To describe properties of the unbounded potential R of a chain X, we introduce the renewal measure on $\mathcal{B}(\mathbb{R})$ by

$$U(B) = \sum_{n\geq 0} \mathsf{P}(S_n \in B) = \sum_{n \geq 0} F^{*n}(B).$$

Here and in the sequel the symbol $*$ stands for the convolution of measures. Note that the finiteness of the mean value $m < 0$ of control variables ξ_n implies [Feller (1966), Chapter 11] that U is locally finite. Moreover, it follows from the elementary renewal theorem that

$$U(-x, \infty) \leq U(E) + kx^+ < \infty \tag{8.17}$$

for all x and some constant $k = k(F)$.

LEMMA 8.2. *Let $m < 0$. Then*

$$0 \leq U(B - x) - R_0(x, B) \leq U(E) < \infty \tag{8.18}$$

for all $x \geq 0$, $B \in \mathcal{B}_+$, so that $\varphi(x) = R_0(x, E) < \infty$ for all $x \geq 0$.

COROLLARY 8.1. *A measure μ on $\mathcal{B}_+$ with $\mu(E) = 0$ belongs to the domain of definition $\mathfrak{M}_0$ of the potential R if and only if the measure $\mu * U$ is finite on $\mathcal{B}_+$. The last condition holds true, in particular, if $\int x\,|\mu|(dx) < \infty$.*

PROOF. Define a kernel

$$\psi(x, B) = \begin{cases} U(B - x) - R_0(x, B), & \text{for } x > 0, \\ U(B - x), & \text{for } x \leq 0. \end{cases}$$

The inequality $\psi(x,B) \leq U(E)$ is evident for $x \leq 0$. It follows from the definitions of R_0 and U in (8.11) and (8.16) that ψ is nonnegative.

The kernel ψ satisfies for $x > 0$ the equation

$$\psi(x,B) = \varphi(x,B) + \mathsf{E}(\psi(\xi + x), 0 < \xi + x \in B) \tag{8.19}$$

that follows from the similar equations for U and R_0, namely, $U = I + U * F$, $R_0 = I + R_0 T$. Here the kernel $\varphi(x,B) = \mathsf{E}(U(B - \xi - x), \xi + x \leq 0)$ is nonnegative and $\mathsf{P}(\xi \in B) = F(B)$. Observe that ψ is the minimal solution of (8.19), since the renewal measure U has the same property and majorises ψ. Iterating (8.19) we get

$$\begin{aligned}\psi(x,B) &= \sum_{t\geq 0} \mathsf{E}(U(B - \xi - S_t - x), S_1 > -x, \ldots, \xi + S_t \leq -x) \\ &= \sum_{n\geq 1} \mathsf{E}(U(B - S_n - x), \tau_x = n, x + S_n \leq 0) \\ &= \sum_{n\geq 1} U(E)\mathsf{P}(\tau_x = n) = U(E),\end{aligned}$$

where ξ does not depend on ξ_t and has the same distribution. The lemma is proved. □

The assertion of the corollary follows from Theorem 5.5 and from (8.18). Indeed, according to (5.15) it suffices to check only the finiteness of the measure μR_0 that, by (8.18), differs from $\mu * U$ in a finite measure. Furthermore, if μ has the finite first moment then the measure $\mu * U$ is finite because of $|\mu * U|(E) \leq \int |\mu|(dx)U(E - x)$ and (8.17).

In relation to Corollary 8.1 consider an induced norm of the perturbation $\delta = G - F$ given by

$$|\delta|_U = \sup_{B\in\mathcal{B}(\mathbb{R})} |\delta * U(B)|. \tag{8.20}$$

THEOREM 8.4. *Let $m < 0$. Then*

$$\sup_{B\in\mathcal{B}_+} \left|\nu(B) - \pi(B) - \cdots - \pi\Delta^{n-1}(B)\right| = O\left(|\delta|_U^n\right), \tag{8.21}$$

$$\sup_{t\geq 0}\sup_{B\in\mathcal{B}_+} |\mathsf{P}_\alpha(Y_t \in B) - \mathsf{P}_\alpha(X_t \in B)| = O(|\delta|_U) \tag{8.22}$$

for all $n \geq 1$ and each initial distribution α of chains X and Y such that $m_\alpha = \int x|\alpha - \pi|(dx) < \infty$, where the kernel Δ is defined in (8.13).

The convergence in (8.21) *and* (8.22) *is uniform with respect to α with finite m_α and to all distributions F, G with bounded $U(E)$ and $k = k(F)$ in* (8.17).

PROOF. In the space $\mathfrak{M}$ of finite measures on $\mathcal{B}_+$, consider a semivariation norm $\|\mu\| = \sup(|\mu(B)|, B \in \mathcal{B}_+)$ that is equivalent to the full variation. As above, the finiteness of $\varphi(x) = R_\varphi(x,E) \leq U(E - x) \leq U(E) + kx$ yields the condition (B) of Theorem 5.5. Hence (8.21) and (8.22) follow from (5.6), (5.43), and (5.69), provided that $\|D\|_0 = O(|\delta|_U)$, where $\|D\|_0 = \|DR_0\|$ according to (5.42).

Using definition (8.9), we obtain for $x \geq 0$, $B \in \mathcal{B}_+$, that

$$DR_0(x,B) = \delta(-\infty,-x]R_0(0,B) + (\delta * U)(B-x) - \int_{-\infty}^{\infty} \delta(dy)\psi(x+y,B), \quad (8.23)$$

where the kernel $\psi(x,B) = U(B-x) - R_0(x,B)$ is bounded uniformly in x because of (8.18). We conclude from (8.23) that

$$\|D\|_0 \leq |\delta|_U + 2U(E)|\delta|(\mathbb{R}), \quad (8.24)$$

where $|\delta|(\mathbb{R}) \leq |\delta|_U$, since the measure $\delta * U$ satisfies the equation $\delta * U - \delta * U * F = \delta$, so that $\||\delta|(\mathbb{R}) \leq |\delta * U|(\mathbb{R})(1 + F(\mathbb{R}))/2 = |\delta|_U$.

We complete the proof of the theorem by referring to (8.24). □

THEOREM 8.5. *Let $m < 0$ and the distribution F^{*k}, for some $k \geq 1$, have an absolutely continuous component.*

Then expansions (8.21) and (8.22) of Theorem 8.4 are valid, where $|\delta|_U$ can be replaced by

$$|\delta|_m = \sup_{B \in \mathcal{B}(\mathbb{R})} \left| \delta(B) + \int_B \delta(x,\infty)\, dx \right|. \quad (8.25)$$

PROOF. Note that

$$\begin{aligned} |\delta * U(B)| &= |(\delta + \delta * U) * (U - U * e)(B)| \\ &\leq \|U - U * e\| \cdot \|\delta + \delta * L\| = \|U - U * e\| \cdot |\delta|_m \end{aligned} \quad (8.26)$$

for $B \in \mathcal{B}(\mathbb{R})$ where L is the Lebesgue measure on $\mathbb{R}_-$ that is extended to $\mathbb{R}_+$ by $L(\mathbb{R}_+) = 0$, the measure e by the definition has the density $\exp\{x\}\mathbb{I}_{x \leq 0}$, and the obvious identity $L = e + e * L$ is used. The finiteness of $\|U - U * e\|$ follows from the nonsingularity of the distribution F, [see Borovkov (1972), Kartashov (1982b)]. □

The necessity of conditions of Theorem 8.5 is established in the following proposition.

THEOREM 8.6. *Let $m < 0$ and $\mathsf{E}\xi_1^2 = \infty$. Assume that for all sufficiently small $|\delta|_m = |G - F|_m$ the distribution G generates an invariant stochastic measure ν of the chain Y such that $\|\nu - \pi\| = O(|\delta|_m)$ as $|\delta|_m \to 0$. Then for some $k \geq 1$ the distribution F^{*k} have an absolutely continuous component.*

PROOF. Consider the family of distributions G of the form $G = (1-\varepsilon)F + \varepsilon H$ where H is an absolutely continuous distribution such that $|H - F|_m < \infty$. Then corresponding transition kernels are of the form $Q = P + \varepsilon B$ where the kernel B is obtained from (8.9) by replacing δ by $H - F$. It is clear that these kernels are bounded on $\mathfrak{M}$. Since $|\delta|_m = |\varepsilon(H-F)|_m = O(\varepsilon)$, conditions of Theorem 5.6 are fulfilled. Thus from Theorem 5.6 we conclude that $\pi B \in \mathfrak{M}_0$. Then by Corollary 8.1 the measure $\pi B * U$ is finite on $\mathcal{B}_+$. Using the definition of B and the boundedness of U on $\mathcal{B}_+$ we prove that the measure

$$\pi * (H - F) * U = \pi * U * H + \pi - \pi * U \quad (8.27)$$

is finite on $\mathcal{B}_+$. The summand $\pi * U * H$ in (8.27) is absolutely continuous and σ-finite. The third term on the right-hand side of (8.27) is infinite on $\mathcal{B}_+$, since

$$\pi * U(E) = \int \pi(dx)U(E-x) \geq \int x\, \pi(dx)|m|^{-1} = \infty$$

because of $\mathsf{E}\,\xi_1^2 = \infty$, see [Borovkov (1972), Chapter 5]. If the difference in (8.27) of infinite measures is finite, then they must have a common infinite component. According to (8.14)

$$\pi * U(B) = \frac{1}{q} \sum_{t \geq 0, n \geq 0} \mathsf{P}(\tau_0 > t, S_{n+t} \in B) \tag{8.28}$$

and therefore the absolutely continuous measure $\pi * U * H$ and at least one of summands in (8.28) have a common component. Then the inequality $F^{*(n+t)}(B) \geq \mathsf{P}(\tau_0 > t, S_{n+t} \in B)$ yields the assertion of the theorem. □

Assuming additional moment restrictions for distributions F and G one can remove the condition on the nonsingularity of F. Define the class of random variables

$$\mathcal{M}_{\lambda\sigma} = \{\xi : \mathsf{E}\,\xi \leq -\lambda, \mathsf{E}\,\xi^2 \leq \sigma(\mathsf{E}\,\xi)^2\}. \tag{8.29}$$

THEOREM 8.7. *Let members of control sequences $\{\xi_n\}$, $\{\eta_n\}$ belong to the class $\mathcal{M}_{\lambda\sigma}$ for some $0 < \lambda < \infty$ and $\sigma < \infty$. Then*

$$\sup_{B \in \mathcal{B}_+} |\nu(B) - \pi(B)| \leq \int_{-\infty}^{\infty} \left(\frac{3\sigma}{2} + \frac{x^+}{\lambda}\right) |G - F|(dx). \tag{8.30}$$

PROOF. We use representation (5.2) of Theorem 5.1 that takes the form $\nu - \pi = \nu\Delta$ in the case under consideration, where the kernel Δ is defined by (8.13). By Lemma 8.2, $R_0(x, B) \leq U(B-x) \leq U(E-x)$ where the function $U(E-x)$ coincides with the renewal function constructed by the sequence $\xi_n' = -\xi_n$. Since $\xi_n \in \mathcal{M}_{\lambda\sigma}$, the Daley inequality in Daley (1980) gives us $U(E-x) \leq x/\lambda + \sigma$ for $x \geq 0$. Therefore,

$$\begin{aligned}
|\nu(B) - \pi(B)| &\leq \int_{[0,\infty)} \nu(dx) \int_{(-x,\infty)} |\delta|(dy) U(E - x - y) \\
&\leq \int_{[0,\infty)} \nu(dx) \int_{(-x,\infty)} |\delta|(dy) \left(\frac{x^+}{\lambda} + \frac{\lambda}{y} + \sigma\right) \\
&\leq \int_{[0,\infty)} \nu(dx) \left(\left(\sigma + \frac{x^+}{\lambda}\right) |\delta|(\mathbb{R}) + \frac{1}{\lambda} \int y^+ |\delta|(dy)\right) \\
&= \int |\delta|(dy) \left(\sigma + \frac{y}{\lambda} + \frac{1}{\lambda} \int x\, \nu(dx)\right),
\end{aligned}$$

for $B \in \mathcal{B}_+$. Here the first moment of the stationary distribution ν constructed by the sequence $\{\eta_n\}$ satisfies the inequality

$$\int x\, \nu(dx) \leq \frac{\mathsf{E}\,\eta_1^2}{2|\mathsf{E}\,\eta_1|} \leq \frac{\sigma\lambda}{2}$$

from Kleinrock (1975), chapter 8.4, provided that $\eta_n \in \mathcal{M}_{\lambda\sigma}$. Substituting this inequality into (8.31), one obtains (8.30). The theorem is proved. □

The kernel R_0 in (8.11) admits also an other representation. Let $\overline{\xi}_t$ be a sequence of lower ladder intervals for ξ_t, [see Prokhorov and Rozanov (1973)], i.e., random variables $\overline{\xi}_t$ are independent identically distributed and $\overline{\xi}_1 = S_{\tau_0}$. Here the moment τ_0 is defined by $\tau_0 = \inf\{t \geq 1 : S_t \leq 0\}$. Denote also by

$$\overline{U}(B) = \sum_{n\geq 0} \mathsf{P}\left(\overline{\xi}_1 + \cdots + \overline{\xi}_n \in B\right)$$

the corresponding renewal measure concentrated on $\mathbb{R}_-$.

LEMMA 8.3. *Let $m < 0$. Then for all $x \geq 0$ and $B \in \mathcal{B}_+$*

$$R_0(x, B) = q \int_{(-x,0]} \pi(B - x - y)\, \overline{U}(dy), \tag{8.32}$$

where $q = 1/\pi(\{0\})$.

PROOF. It is clear that $\tau_0 \leq \tau_x$. We break up all events under the probability sign in definition (8.11) into two parts depending on $\tau_0 \leq t < \tau_x$ or $t < \tau_0$. The sum of probabilities of the events of the first kind is equal to $\mathsf{E}(R_0(x + S_{\tau_0}, B), -x < S_{\tau_0} \leq 0)$ and one of the events of the second kind is equal to $q\pi(B - x)$ because of (8.14). For this reason, the renewal equation $R_0(x, B) = q\pi(B - x) + \mathsf{E}(R_0(x + \overline{\xi}_1, B), -x < \overline{\xi}_1 \leq 0)$ is satisfied and (8.32) is proved. □

REMARK 8.1. One can derive from (8.32) that the assertion of Theorem 8.4 remains true if the measure U in (8.20) is replaced by the measure $\overline{U}$ (see representation (8.37) for the kernel Δ).

Using Lemma 8.3 one can also obtain an analogue of Theorem 8.4 for the uniform Kolmogorov norm

$$\|\mu\|_r = \sup_{x\geq 0} |\mu(x, \infty)|. \tag{8.33}$$

THEOREM 8.8. *Let $m < 0$, $\mathsf{E}\eta_n < 0$, and let the distribution F^{*k} have an absolutely continuous component. Then*

$$\sup_{x\geq 0} |\nu(x, \infty) - \pi(x, \infty)| = O(|\delta|_r), \tag{8.34}$$

$$\sup_{x\geq 0} |\nu(x, \infty) - \pi(x, \infty) - \pi\Delta(x, \infty)| = O(|\delta|_r^2 + |\delta|_{rr}), \tag{8.35}$$

where the kernel Δ is defined by (8.13), $\delta = G - F$, and by the definition

$$\begin{aligned} |\delta|_r &= \sup_{x\in\mathbb{R}} |\delta(x, \infty)| + \int_{\mathbb{R}} |\delta(x, \infty)|\, dx, \\ |\delta|_{rr} &= \sup_{x,y\in\mathbb{R}} \left| \int_{(-x,\infty)} \delta(dz)\delta(y - x - z, y) \right|. \end{aligned} \tag{8.36}$$

PROOF. Substituting (8.32) into (8.13) we conclude that

$$\Delta(x,B) = q\int \pi(dy)\left(\mu(B-y)^{+} - x\right) - q\pi(B)\mu(E-x) \tag{8.37}$$

for $x \geq 0$ and $B \in \mathcal{B}_{+}$, where $\mu = \delta * \overline{U}$ and $B^{+} = B \cap E$.

Denoting by $\alpha(z) = \alpha(z,\infty)$ the tail distribution for any measure $\alpha(\cdot)$ and substituting $B = (z,\infty)$ into (8.37) we get

$$\Delta(x,z) = q\int_0^z \pi(dv)\mu(z-x-v). \tag{8.38}$$

According to (5.2) and the definition of $\Delta = DR$ we obtain $(\nu - \pi)(z) = \nu\Delta(z) = \int \nu(dx)\Delta(x,z)$. Now from (8.38) we conclude that

$$||\nu - \pi||_r \leq q||\mu||_r = q||\delta * \overline{U}||_r. \tag{8.39}$$

As in the proof of Theorem 8.5, applying the first inequality in (8.26), it is easy to estimate the right-hand side of (8.39) as follows

$$\begin{aligned} \sup_{x\in\mathbb{R}} \left|\delta * \overline{U}(x)\right| &= \sup_{x\in\mathbb{R}} \left|(\overline{U} - \overline{U} * e) * (\delta + \delta * L)(x)\right| \\ &\leq \left|\overline{U} - \overline{U} * e\right|(\mathbb{R})||\delta + \delta * U||_r = \overline{M}|\delta|_r, \end{aligned} \tag{8.40}$$

where $\overline{M} < \infty$, since F is nonsingular. Estimates (8.40) and (8.39) immediately imply (8.34).

To obtain (8.35) we evaluate from (8.37) and (8.38) the square of the kernel Δ as follows

$$\begin{aligned} \Delta^2(x,z) &= q^2\int \pi(du)\int_{(-x,\infty)} \mu(dy)\int_{[0,z]} \pi(dv)\mu(z-v-u-x-y) \\ &\quad - q^2\int \pi(du)\int_{(-x,\infty)} \mu(dy)\int_{[0,z]} \pi(dv)\mu(z-u-v) \\ &= q^2\int \pi(du)\int_{(0,z)} \pi(dv)\int_{(-x,\infty)} \mu(dy)\mu(z-v-u-x-y, z-u-v) \\ &= q^2\int \pi(du)\int_{[0,z]} \pi(dv)\mu\otimes\mu\left(\Gamma_{x,z-u-v}\right), \mu(z-v-u-x-y) \end{aligned} \tag{8.41}$$

where $\Gamma_{a,b} = \{(s,t): s \leq b < s+t+a\}$, the measure $\mu = \delta * \overline{U}$ is defined as above, and the symbol $\otimes$ stands for the direct product of measures.

According to identity (5.3) of Theorem 5.1, the representation $\nu(z) = \pi(z) + \pi\Delta(z) + \nu\Delta(z)$ holds true. Then from (8.41) we conclude that

$$||\nu - \pi - \pi\Delta||_r \leq q^2 \sup_{x,y\in\mathbb{R}} |\mu\otimes\mu(\Gamma_{x,y})|. \tag{8.42}$$

Analogously to (8.40), one can transform the measure $\mu = \delta * \overline{U}$ to the form $\mu = \omega * \gamma$, where $\omega = \delta + \delta * L$ and $\gamma = \overline{U} - e * \overline{U}$ has a bounded variation provided F is nonsingular. Now we estimate the right hand side of (8.42) as follows

$$\begin{aligned} |\mu \otimes \mu\,(\Gamma_{x,y})| &= \left| \int \gamma(dy) \int \gamma(dv) \omega \otimes \omega\,(\Gamma_{x+v,y-u}) \right| \\ &\leq \left(|\gamma|(\mathbb{R})\right)^2 \sup_{x,y} |\omega \otimes \omega(\Gamma_{x,y})| = O\left(|\delta|_r^2 + |\delta|_{rr}\right), \end{aligned} \tag{8.43}$$

since $|\omega \otimes \omega(\Gamma_{x,y})| \leq |\delta \otimes \delta(\Gamma_{x,y})| + 2\,|(\delta * L) * \delta(\Gamma_{x,y})|$. The upper bounds of the first summand is $O(|\delta|_r^2)$ and the one of the second one is $O(|\delta|_{rr})$. The theorem is proved. □

CHAPTER IX

INEQUALITIES IN RENYI'S THEOREM

In this Chapter we consider sums of a random geometric number of independent identically distributed random variables. Such sums are used in reliability theory to describe the time to a failure of a renewable system. We interpret the renewal intervals in the system as summands as well as the renewal amount before a failure as its number. The classical Renyi's theorem states that the distributions of sums under consideration are approximated by an exponential distribution if the probability of a failure over a renewal period converges to zero.

A corresponding limit theorem was proved by Solov′ev (1971). In Kalashnikov (1983) and Kalashnikov and Vsekhsvyatski (1989) various estimates of convergence in Renyi's theorem are given under the assumption that the second moment is finite. We improve some of their results using methods of Chapter 7.

9.1. Main results

Let $(\xi_t, t \geq 0)$ be a sequence of nonnegative independent identically distributed random variables with the common distribution $F(B) = \mathsf{P}(\xi_t \in B)$ on the Borel σ-field $\mathfrak{B}(\mathbb{R}_+)$ and with moments

$$\mathsf{E}\,\xi_0 = 1, \qquad \mathsf{E}\,\xi_0^2 = 2\varkappa < \infty. \tag{9.1}$$

We introduce the renewal measure

$$U(B) = \sum_{t\geq 0} F^{*t}(B),$$

the renewal function $U(x) = U([0,x])$, and the functions

$$p(x) = \mathsf{P}(\xi_0 > x), \qquad \overline{p}(x) = \int_{(x,\infty)} p(u)\,du.$$

Put $S(n) = \xi_0 + \xi_1 + \cdots + \xi_n$, $n \geq 0$.

Let ζ be a discrete random variable independent of (ξ_t) and with a geometric distribution $\mathsf{P}(\zeta = n) = \theta(1-\theta)^n$, $n \geq 0$. Obviously, $\zeta \to \infty$ and $S(\zeta) \to \infty$ as $\theta \to 0$. The asymptotic behaviour of the sums $S(\zeta)$ is estimated in the following theorem.

THEOREM 9.1. *Let summands have finite second moments and* (9.1) *hold true. Then*

$$\sup_{x\geq 0} |\mathsf{P}(qS(\zeta) > x - \exp\{-x\}| \leq 4q \sup_{x\geq 0} |\delta * U(x)| \tag{9.2}$$

$$\leq 4\frac{\theta\varkappa}{1-\theta+\theta\varkappa}, \tag{9.3}$$

where the normalization term is $q = \theta/(1-\theta+\theta\varkappa)$, $\delta(x) = \overline{p}(x) - \varkappa p(x)$, and $*$ denotes the convolution of a function and the renewal measure.

REMARK 9.1. The right-hand side of (9.2) tends to zero not only as $\theta \to 0$, but also as the common distribution of summands is close to an exponential distribution. In this case the function δ tends to zero.

COROLLARY 9.1. *Suppose that the distribution F varies in a series scheme in such a way that the corresponding family of distributions is uniformly nonsingular (this condition is fulfilled, for example, if all distributions have a fixed common nonsingular component; for other criteria see Kartashov (1982)). Then there is a constant C such that*

$$\sup_{x\geq 0} |\mathsf{P}(qS(\zeta) > x) - \exp\{-x\}| \leq C\theta|\delta|_r, \tag{9.4}$$

where

$$|\delta|_r = \sup_{x\geq 0}|\delta(x)| + \int_{(0,\infty)} |\delta(x)|\, dx,$$

and the function $\delta = \overline{p} - \varkappa p$ is defined above. For example, C can be taken to be $8\sup_F \operatorname{var}(U - e * U)$ where e is the exponential density with parameter 1.

The estimate (9.3) can be refined somewhat by using another normalization.

THEOREM 9.2. *Let (9.1) be fulfilled. Then for all $\theta < 1$*

$$\sup_{x\geq 0} |\mathsf{P}(rS(\zeta) > x) - \exp\{-x\}| \leq 2\frac{\theta\varkappa}{1-\theta}, \tag{9.5}$$

where the normalization term is $r = \theta/(1-\theta)$.

REMARK 9.2. The absolute constant 2 on the right-hand side of (9.5) cannot be decreased by a factor of more than 2 because the upper limit as $\theta \to 0$ of the left-hand side of (9.5) divided by θ is not less than $|\varkappa - 1|$. Moreover, the validity of (9.5) with $O(\theta)$ as $\theta \to 0$ on the right-hand side implies the finiteness of the second moment $\varkappa = \mathsf{E}\,\xi_0^2/2$.

REMARK 9.3. Inequalities (9.3) and (9.5) are valid also in the case when the distribution of the last (or of the first) summand ξ_0' in the sum $S'(\zeta)$ differs from the others. To make the estimate correct one can add to the right-hand sides of these inequalities $\max(\theta, qm)$ and $\max(\theta, rm)$ respectively, where $m = \mathsf{E}\,\xi_0'$.

A stationary modification of the sums $S'(\zeta)$ admits an essentially better approximation.

THEOREM 9.3. *Suppose that the moment $\varkappa_3 = \mathsf{E}\,\xi_1^3/6$ is finite, and the distribution of the first summand is $\mathsf{P}\,(\xi_0' < x) = 1 - \overline{p}(x)$. Then*

$$\sup_{x\geq 0} |\mathsf{P}\,(rS'(\zeta) > x) - \exp\{-x\}(1 + \beta + x(r\varkappa - \beta))| \leq r^2(2\varkappa_3 + 4\varkappa^2), \tag{9.6}$$

where $r = \theta/(1-\theta)$ and $\beta = r^2\varkappa_3$.

9.2. Discrete summands

Inequalities of Theorems 9.1–9.3 will be derived from their discrete analogues. We shall formulate and prove them.

Assume first that the random variables ξ_t^d, $t \geq 0$, take only positive integer values, have a common discrete distribution $g = (g_i, i \geq 1)$, and have moments

$$\lambda_d = \mathsf{E}\,\xi_0^d, \qquad \varkappa_d = \mathsf{E}\,\xi_0^d \frac{\xi_0^d + 1}{2\lambda_d^2}. \tag{9.7}$$

We introduce a renewal sequence

$$U_t^d = 1 + \sum_{l \geq 0} \mathsf{P}(S_d(l) \leq t)$$

connected with g, where $S_d(l) = \xi_0^d + \cdots + \xi_l^d$ is the corresponding sequence of sums.

Let ζ be the same as above.

THEOREM 9.4. *If moments* (9.7) *are finite then for each* $s \leq \infty$

$$\sup_{t \leq s} \left| \mathsf{P}(S_d(\zeta) > t) - \left(1 - \frac{q_d}{\lambda_d}\right)^t \right| \leq 4q_d \sup_{t \leq s} \left| U_t^d - \frac{t}{\lambda_d} - \varkappa_d \right|, \tag{9.8}$$

$$\sup_{t \leq s} \left| \mathsf{P}(S_d(\zeta) > t) - \left(1 - \frac{r_d}{\lambda_d}\right)^t \right| \leq r_d \sup_{t \leq s} \left| U_t^d - \frac{t+1}{\lambda_d} \right|, \tag{9.9}$$

where $q_d = \theta/(1 - \theta + \theta\varkappa_d)$ *and* $r_d = \theta/(1 - \theta + \theta/\lambda_d)$.

PROOF OF THEOREM 9.4. To make reading easier the index d will be omitted. Inequalities (9.8) and (9.9) will be derived from the estimate (7.24) of Theorem 7.3 which contains explicit bounds of asymptotics of stopping times on a homogeneous Markov chain. To apply Theorem 7.3 we consider the jump process

$$X_t = \inf(S(l) - t - 1, S(l) > t). \tag{9.10}$$

It is well known that this process forms a Markov chain on the space $E = \{0, 1, \dots\}$, satisfies the recurrent relations

$$X_{t+1} = X_t I(X_t > 0) + \xi_{t+1} I(X_t = 0) - 1,$$

and is described by the transition probability matrix $P = (P_{ij})$ where

$$P_{ij} = \delta_{0i} g_{j+1} + \delta_{i,j+1}. \tag{9.11}$$

Here and below δ_{ij} is the Kronecker symbol. The chain X_t has a unique invariant (stationary) distribution $\pi = (\pi_j)$,

$$\pi_j = \frac{1}{\lambda} \sum_{i > j} g_i,$$

and its initial distribution α coincides with the shifted distribution g so that $X_0 = \xi_0 - 1$.

On an extension of the original probability space we consider another Markov chain $Z_t = (X_t, Y_t)$, where the auxiliary process is $Y_t = \beta(\gamma(t) - 1)$, the local time is $\gamma(t) = \sum_{0 \le s < t} I(X_s = 0)$, and the random variables $(\beta(n), n \ge 0)$ are independent of one another and independent of X_t, have a common distribution $\mathsf{P}(\beta(n) = 1) = \theta = 1 - \mathsf{P}(\beta(n) = 0)$, and by definition $\gamma(0) = 0$ and $\beta(-1) = 0$. It is clear that the chain X_t is adapted to the flow $\mathfrak{F}_t = \sigma[Z_s, s \le t]$.

We define the first passage time

$$\tau = \inf(t \ge 1 \colon Y_t = 1).$$

Obviously, τ is the homogeneous killing moment for $(X_t, \mathfrak{F}_t)$, so the equality (7.1) of Chapter 7 holds.

From (9.10) and from the relations

$$\{Y_{t-1} \ne Y_t\} \subset \{X_{t-1} = 0\} = \bigcup_{n \ge 0} \{S(n) = t\}$$

we conclude that

$$\begin{aligned}\{\tau = t\} &= \{\tau = t, Y_t = 1, X_{t-1} = 0\} = \bigcup_{n \ge 0} \{\tau = t, S(n) = t, \beta(n) = 1\} \\ &= \bigcup_{n \ge 0} \{\beta(0) = 0, \dots, \beta(n-1) = 0,\ \beta(n) = 1,\ S(n) = t\} = \{S(\zeta) = t\}.\end{aligned}$$

Here the time

$$\zeta = \inf(n \ge 0 \colon \beta(n) = 1)$$

does not depend on X_t and has the geometric distribution with parameter θ.

Therefore the results of Theorem 7.3 are applicable to the sum $S(\zeta) = \tau$. We remark that by (7.4) the distribution of the moment τ is given by powers of the substochastic matrix $Q = (Q_{ij})$ where

$$Q_{ij} = P_i(X_1 = j, \tau > 1) = (1 - \theta)\delta_{0i} g_{j+1} + \delta_{i,j+1}. \tag{9.12}$$

As the space $\mathfrak{M}$ of measures on E we choose the set of all measures (summable sequences) $\mu = (\mu_j)$ with finite norm $\|\mu\| = \sum (j+1)|\mu_j|$. The space $\mathfrak{N}$ dual to $\mathfrak{M}$ consists of the functions (finite valued sequences) $f = (f_i)$ on E with at most linear growth. We remark that the spaces $\mathfrak{M}$ and $\mathfrak{N}$ are invariant for P and Q.

Using (9.12), we conclude that, in the case under consideration, equation (7.7), for the function $h = (h_i)$ defined as in Theorem 7.1, has the following form:

$$\begin{aligned} h_0 &= (1 - \theta) \sum g_{j+1} h_j + \rho - 1 + \theta, \\ h_i &= h_{i-1} + \rho - 1, \qquad i \ge 1. \end{aligned} \tag{9.13}$$

Setting $p = 1 - q/\lambda$ and solving (9.13) we see that the function

$$h_i = q\left(\varkappa - \frac{i+1}{\lambda}\right) \tag{9.14}$$

belongs to the space $\mathfrak{N}$ and satisfies the conditions of Theorem 7.3.

Using the second part of inequality (7.24) and Remarks 7.1 and 7.4, we conclude that the left-hand side of (9.8) does not exceed the quantity

$$2\varepsilon'_\alpha = 2\sup_{t\le s}\left|\alpha Q^t h\right|, \tag{9.15}$$

where α is the initial distribution of the chain X, and the matrix Q and the vector h are computed in (9.12) and (9.14). Note that in a discrete space the product of a measure and a kernel, the product of a kernel and a function reduce to the multiplications of row vectors to a matrix from the left and of matrix to a column vector from the right, respectively.

Computing the difference $P^t_{ij} - Q^t_{ij} = \mathsf{P}_i(X_t = j, \tau \le t)$ and using the formula for the total probability, the Markov property (7.1) of the time τ, and also equality (9.12), we obtain the identity

$$Q^t_{ij} = P^t_{ij} - \sum_{l=1}^{t} Q^{l-1}_{i0}\theta\alpha P^{t-l}_j. \tag{9.16}$$

Multiplying (9.16) by $\alpha_i = g_{i+1}$ and by h_j and summing over all $i, j \ge 0$, we deduce that the deviation on the left-hand side of (9.8) does not exceed

$$2\varepsilon'_\alpha \le 2\sup_{t\le s}\left|\sum_{i,j}\alpha_i P^t_{ij}h_j\right|\left(1+\theta\sum_{i,t}\alpha_i Q^t_{i0}\right) \le 4\sup_{t\le s}\left|\alpha P^t h\right|, \tag{9.17}$$

since

$$\begin{aligned}\theta\sum_{i,t}\alpha_i Q^t_{i0} &= \theta\sum_{l,n}\mathsf{P}(S(n) = l+1)(1-\theta)^n \\ &= \theta\sum_{n\ge 0}\mathsf{P}\Big(\bigcup_{l\ge 0}\{S(n) = l+1\}\Big)(1-\theta)^n \le 1,\end{aligned}$$

where $S(n) = \xi_0 + \cdots + \xi_n$ is the discrete random walk introduced above.

To prove (9.8) it remains to see that, according to (9.10) and (9.14),

$$\alpha P^t h = q\,\mathsf{E}\left(\varkappa - \frac{X_t+1}{\lambda}\right) = q\left(\varkappa - \frac{\lambda U_t - t}{\lambda}\right) = -q\left(U_t - \frac{t}{\lambda} - \varkappa\right) \tag{9.18}$$

in view of properties of the process (9.10).

Substituting (9.18) into (9.17), we get (9.8).

The estimate (9.9) is derived according to the same scheme: setting $\rho = 1 - r/\lambda$ and solving system (7.7) in Chapter 7, we find that $h_i = ri/\lambda$, $i \ge 0$. Note that the function h_i is nonnegative. In this case, according to the last assertion of the Remark 7.3, the left-hand side of inequality (7.24) does not exceed ε'_α without the additional factor of 2. For the same reason, $\varepsilon'_\alpha = \sup \alpha Q^t h \le \sup \alpha P^t h$, because $Q \le P$. Estimating the last expression as in (9.18) we arrive at (9.9).

9.3. General summands

Consider the sequence of discrete variables $\xi_t^d = 1 + [n\xi_t]$. Let $S_d(\lambda)$, λ_d, $\varkappa_d$, q_d, r_d, and U_t^d be constructed from ξ_t^d as in Theorem 9.4.

Note that $n^{-1}\xi_t^d \to \xi_t$ as $n \to \infty$ for all t and ω, and $n^{-1}S_d(t) \to S(t)$, $q_d \to q$, $r_d \to r$, $\varkappa_d \to \varkappa$, $\lambda_d \sim \lambda$ as $n \to \infty$. Moreover, since the series

$$U_{nx}^d = \sum_{t\geq 0} \mathsf{P}(S_d(t) \leq nx) \leq U(x)$$

converges uniformly with respect to x, it follows that $U_{nx}^d \to U(x)$ as $n \to \infty$ almost surely with respect to x in each interval of variation of x.

We write inequality (9.8) for the discrete random variables we have introduced, substituting in place of the left-hand side its value at the point $s = nx$, and replacing the index $t \leq s$ on the right-hand side by ny, $y \leq x$. Since $S_d(\zeta)/n \to S(\zeta)$ as $n \to \infty$ for all ω and $(1 - q_d/\lambda_d)^{nx} \to \exp\{-qx\}$, we get in the limit as $n \to \infty$ the inequality

$$\left|\mathsf{P}(S(\zeta) > x) - \exp\{-qx\}\right| \leq 4q \sup_{y\leq x} |U(y) - y - \varkappa|. \tag{9.19}$$

Extending the domain of variation of y on the right-hand side of (9.19) and replacing x by x/q on the left-hand side, we arrive at the inequality (9.2), since

$$\begin{aligned}\delta * U(x) &= \overline{p} * U(x) - \varkappa p * U(x) = (\mathbb{I} - \mathbb{I} * p) * U(x) - \varkappa \\ &= \mathbb{I} * U(x) - \mathbb{I} * \mathbb{I} - \varkappa = U(x) - x - \varkappa\end{aligned} \tag{9.20}$$

by properties of the renewal measure. Here we use the known identity $p * U(x) = 1$. Inequality (9.2) is proved.

To derive (9.3) it suffices to note that $U(x) \geq x$ and $U(x) \leq x + 2\varkappa$ by a Daley (1980) inequality. This by (9.20) gives us that

$$|\delta * U(x)| = |U(x) - x - \varkappa| \leq \max(-\varkappa, \varkappa) = \varkappa. \tag{9.21}$$

Inequality (9.4) follows from the identity

$$\delta * U = (\delta + \delta * L) * (U - e * U), \tag{9.22}$$

where L is the Lebesgue measure. The uniform boundedness of the total variation of the measure $U - e * U$ follows from the uniform renewal theorem by Kartashov (1982a,b).

Theorem 9.2 can be proved similarly: the inequality

$$\left|\mathsf{P}(S(\zeta) > x) - \exp\{-rx\}\right| \leq r \sup_{y\leq x} [U(y) - y] \tag{9.23}$$

is first derived from (9.9), and then the Daley (1980) inequality is used.

To prove Remark 9.2, we let C be the limit superior indicated in it. Then

$$|\mathsf{P}(S(\zeta) \leq x) - 1 + \exp\{-rx\}| \leq C\theta + o(\theta) \tag{9.24}$$

uniformly with respect to $x \geq 0$ as $\theta \to 0$. Note that

$$\lim_{\theta\to 0}\frac{1}{\theta}\mathsf{P}(S(\zeta)\leq x)=\lim_{\theta\to 0}\sum_{n\geq 0}(1-\theta)^n\mathsf{P}(S(n)\leq x)=\sum_{n\geq 0}\mathsf{P}(S(n)\leq x)=U(x)-1,$$

$$\lim_{\theta\to 0}\frac{1}{\theta}[1-\exp\{-rx\}]=\lim_{\theta\to 0}\frac{rx}{\theta}=x.$$

Consequently, it follows from (9.24) that $|U(x)-1-x|\leq C$ for all $x\geq 0$. However, $U(x)-x\to\infty$ as $x\to\infty$ if $\mathsf{E}\,\xi_0^2=\infty$, [see Borovkov (1972), Appendix 4], and $U(x)-x\to\varkappa$ as $x\to\infty$ if $\varkappa<\infty$. Hence, under the conditions of Remark 9.2, $\varkappa<\infty$ and $|\varkappa-1|\leq C$. It follows from (9.5) that $C\leq 2\varkappa$. Therefore Remark 9.2 is valid.

Validity of the Remark 9.3 is based on the identity

$$\widetilde{\Delta}=\theta(\widetilde{G}-e)+(1-\theta)e*\widetilde{G}+(1-\theta)\Delta*\widetilde{F}, \tag{9.25}$$

where Δ is the deviation on the left-hand sides of (9.2) and (9.4) inside the absolute value sign, $\widetilde{\Delta}$ is its analogue after replacing ξ_0 by $\tilde{\xi}_0$, e is the exponential density and the exponential tail, and $\widetilde{F}$ and $\widetilde{G}$ are the distribution and its tail for the random variable $q\tilde{\xi}_0$. Equality (9.25) can be derived from the coincidence of the distribution of the variables $\widetilde{S}(\zeta)$ and $\tilde{\xi}_0+\chi S(\zeta)$, where $\tilde{\xi}_0$ and $S(\zeta)$ are independent, and χ is the indicator function of an event of probability $1-\theta$. The value $\max(\theta,\widetilde{qm})$ mentioned in the remark is a uniform estimate of the first two terms on the right-hand side of (9.25), because $e\leq 1$ and $\widetilde{G}-e\leq e*\widetilde{G}$.

The proof of Theorem 9.3 is analogous to the preceding. It is based on inequality (7.25) of Theorem 7.3 that implies a discrete-time analogue of (9.6). To obtain second order approximation we use the following representation for the discrete variant of the deviation on the left-hand side of (9.6)

$$\pi[Q]\mathbb{I}-G+(\nu h)G*(\delta-g)=(\delta-g)*(\delta-g)*\nu[Q]h. \tag{9.26}$$

This representation is obtained in the proof of inequality (7.25) of Theorem 7.3. Here the measure $\nu=(\nu_j)$ and the function $h=(h_i)$ are solutions to (7.6) and (7.7) of the form $\nu_j=-r_d\overline{G}_j/\lambda_d$ and $h_i=r_d i/\lambda_d$, respectively, where $r_d=\theta/(1-\theta+\theta/\lambda_d)$ and $\overline{G}_j=\sum_{i>j}\sum_{k>i}g_k$. The parameter of the geometric distribution in (7.6) and (7.7) is $\rho=1-r_d/\lambda_d$. The condition $\varkappa_3<\infty$ implies that ν and h belong to the spaces $\mathfrak{M}$ and $\mathfrak{N}$ respectively that are introduced above in the proof of Theorem 9.4. As above, from (9.26) we derive that the discrete variant of the deviation on left-hand side of (9.6) does not exceed the quantity

$$|(\delta-g)(\delta-g)|_1|\nu[Q]h|_0\leq 4\sup_{t\geq 0}\left|\nu Q^t h\right|. \tag{9.27}$$

The right-hand side of (9.27) can be estimated in the same way as in the proof of Theorem 9.4 by using identity (9.18). The transition from the discrete-time summands to the general case is realized according to the same scheme.

CHAPTER X

BIBLIOGRAPHICAL NOTES

Basic notions of the theory of linear operators on cones are introduced by Bogolyubov and Krein (1946), Krein and Rootman (1948), Birkhof (1967), Krein et al. (1972).

The theory of quasicompact chains for the total variation norm is well developed by Doeblin (1940), Dunford and Schvartz (1953), Goldenstein, Gohberg, and Markus (1957), Neveu (1964), Revuz (1975), Fortet (1978), Korolyuk and Turbin (1978), Turbin (1981). This theory is widely used in proofs of limit theorems such as central limit theorem, consolidation theorems, and related problems, see Doob (1953), Nagaev (1957), Davydov (1973), Lifshits (1978), Zubkov (1979), Kovalenko (1977), Kalashnikov (1981), Anisimov (1988), Korolyuk and Turbin (1976)–(1982), Korolyuk (1989). Note that, according to Example 4.2, a simple discrete random walk on the half-line is not quasicompact but it is uniformly ergodic under an appropriate choice of a norm, see Kartashov (1977)-(1984).

The classical minorization condition for transition kernels is considered by Doeblin (1940) and is used by Sevast'anov (1957), Nagaev (1961b, 1982, 1989) and systematically investigated also in Orey (1971), Revuz (1975), and Athreya and Ney (1978), Tweedie and Nummelin (1978). Test functions are used by Kalashnikov (1978) to obtain ergodicity and stability conditions for Markov processes.

An approach closed to the one introduced in Chapter 1 is proposed by Gudinas (1982a, 1982b), Lifshits (1978) for the proof of a central limit theorem. Tweedie (1983) independently considers uniform ergodic chains in a weighted variation norm similar to that in Chapter 1. Sufficient conditions for the exponential ergodicity are given also by Spieksma and Tweedie (1994).

Some explicit estimates for the ergodicity of homogeneous and inhomogeneous finite Markov chains in various norms are obtained by Seneta (1979, 1993b, 1993c), Tan (1982, 1983), Yadrenko (1988), Lesanovsky (1990), Cohen, Imasa, Rautu, Ruskai, Seneta, and Zbaganu (1993). An analogue to Theorem 2.3 for the total variation norm is given by Zubkov (1979). A relationship between spectral properties of a transition operator and the exponential convergence to the stationarity is investigated by Diaconis and Strook (1991), Fill (1991), Chen (1991).

Asymptotic representations in ergodic theorems for Markov chains, semi–Markov processes, and queuing processes are derived in Kartashov (1977, 1985a, 1985d, 1985f). Some closed problems are considered earlier in Goldenstein, Gohberg, and Markus (1957). General conditions for the exponential asymptotics in ergodic theorems are given by Nummelin and Tweedie (1976, 1978), Popov (1979), Tweedie (1981), Nummelin and Tuominen (1982).

Basic results of the perturbation theory are developed by Kato (1966).

The general notion of the quantitative stability of a stochastic sequence is introduced by Zolotarev (1975, 1976a, 1976b, 1986), Borovkov (1977, 1988, 1990, 1991a, 1991b), Kalashnikov (1977, 1978, 1980, 1981, 1982, 1983, 1989). These authors develop various methods to investigate the stability such as the method of metric distances, the method of renewating events, the method of proximity points. The notion of strong stable chains with respect to a norm is introduced by Kartashov (1981a, 1981b, 1984, 1985c).

The weak stability of distributions of Markov processes with values in abstract spaces is proved in Tweedie (1975a), Franken and Stoyan (1974), Stoyan (1977), Wolf (1980). Criteria for weak stability of Markov chains and Markov processes with values in a topological space are obtained by Meyn (1991), Meyn and Tweedie (1992, 1993).

The qualitative stability and continuity of denumerable Markov chains and related questions are investigated by Filonov (1978), Malyshev and Menshikov (1979), Plocki (1979), Heyman (1991). Estimates of the stability under perturbations of a special form are given by Hartfiel (1981), Meyer (1980), Tweedie (1971, 1975b, 1980), Fonderic and Meyer (1986). The result of Example 5.1 improves a corresponding estimate by Tweedie (1980).

A series of topological estimates for a invariant vector of a finite Markov matrix is obtained by Hartfiel (1994). Hunter (1991) proposes a computational algorithm to obtain the ergodic distribution of a perturbed chain. Seneta (1988, 1993a, 1993d) proves explicit estimates for the stability of an invariant vector of a finite Markov chain in terms of special ergodicity coefficients.

A general notion of the potential of an ergodic chain is introduced by Neveu (1964b, 1972), Orey (1964), Revuz (1975), Plotkin and Turbin (1975), Kemeny, Snell, and Knepp (1976), Korolyuk and Shurenkov (1977), Turbin (1981). A relation between a generalized potential and a potential of a residual kernel is investigated by Shurenkov (1986). Syski (1992) develops the probabilistic methods in the potential theory. Explicit estimates for a norm of a potential of a finite chain are given by Meyer (1993), Seneta (1993d).

The problem of states consolidation for a Markov chain taking values in an abstract space is proposed and investigated by Korolyuk and Turbin (1976). They used the perturbation theory of linear operators. Other methods are developed by Anisimov (1972), Kovalenko (1980). Methods of the potential theory for this problem are developed in Kartashov (1985b, 1986). Estimates of convergence for countable chains are given earlier by Anisimov (1980), Kalashnikov (1981), Bobrova (1983).

Numerical approximations for the stationary distribution of nearly uncoupled finite Markov chains are proposed and investigated by Louchard and Latouche (1990), Stewart (1991), Barlow (1993), O'Cinneide (1993).

An analogue to Corollary 7.4 is obtained by Zubkov (1979) in terms of mixing coefficients. Another approach based on the regeneration theory is proposed earlier by Solov'ev (1971) and developed also by Kovalenko (1973, 1977, 1980). A survey on rare events results in queuing systems is given by Kovalenko (1994). A special case of a finite chain under a stationary initial distribution is considered by Aldous (1982). Applications of Renyi's theorem are given by Kartashov (1991). An another approach to this problem is proposed earlier by Kalashnikov (1983), Kalashnikov and Vsekhsvyatski (1989). Estimates of convergence of thinning renewal processes are obtained by Sugakova (1990, 1991, 1993).

The stability of queuing systems was earlier investigated by Kennedy (1972), Franken and Stoyan (1974), Zolotarev (1975, 1976b), Borovkov (1977), Kalashnikov (1977, 1982), Ahmarov (1981). A similar result in Tweedie (1983) is an evident corollary of Theorems 8.1 and 8.2. Uniform renewal theorems used in Chapter 8 are proved by Kartashov (1982a, 1982b). Daley, Kreinin, and Trengove (1992) gave a survey on inequalities for the waiting-time in a single-server queue.

Another examples for the strong stability results in queuing are given by Aissani and Kartashov (1983b), Aissani (1988). An exponential decay in the ergodic theorem for queuing systems is established earlier by Neuts and Teugels (1975).

Kozarovitsky (1991) proves the uniform ergodicity and the strong stability theorems for general valued semi-Markov processes. Komashko (1990) investigates the generalized potential for special classes of Markov chains appearing in the queuing theory.

Notions of the uniform ergodicity and stability of nonhomogeneous Markov chains with a continuous time are proposed and investigated by Zeifman (1985, 1992).

References

1. I. Ahmarov, *Speed of convergence in ergodicity theorems and stability in systems with refusals*, Teor. Veroyatnost. i Primenen. **26** (1981), 182–189; English transl. in Theory Probab. Appl. **26** (1981).
2. D. Aissani, *Perturbation of transition operators for the study of the stability of Markov chains*, Cahiers Math. **1** (1988), 7–12. (French)
3. D. Aissani and N. V. Kartashov, *Ergodicity and stability of Markov chains with respect to operator topologies in space of transition kernels*, Doklady AN USSR, ser. A **11** (1983a), 3–5. (Russian)
4. ———, *A strong stability of the embedded Markov chain in the M/G/1 queuing system*, Teor. Veroyatnost. i Mat. Statist. **29** (1983b), 3–7; English transl. in Theory Probab. Math. Statist. **29** (1984).
5. D. J. Aldous, *On d-convergence of Markov chains*, Math. Proc. Cambridge Phil. Soc. **2** (1981), 331–333.
6. ———, *Markov chains with almost exponential hitting times*, Stoch. Proc. Appl. **13** (1982), 305–310.
7. V. V. Anisimov, *On the limiting behavior of a semi-Markov process with splitting set of states*, Doklady AN SSSR **206** (1972), 3–5. (Russian)
8. ———, *Limit theorems for processes that admit an asymptotic consolidation of a component*, Teor. Veroyatnost. i Mat. Statist. **22** (1980), 3–15; English transl. in Theory Probab. Math. Statist. **22** (1981).
9. ———, *Stochastic processes with a discrete component*, "Vyscha shkola", Kiev, 1988. (Russian)
10. K. B. Athreya and P. Ney, *A new approach to the limit theory of recurrent Markov chains*, Trans. Amer. Math. Soc. **245** (1978), 493–501.
11. A. Baltrunas, *Ergodicity of a class of Markov chains*, Litovsk. Math. Sbornik **4** (1979), 3–11. (Russian)
12. J. L. Barlow, *Perturbation results for nearly uncoupled Markov chains with applications to iterative methods*, Numer. Math. **65** (1993), 51–62.
13. G. Birkhoff, *Lattice theory*, American Mathematical Society, Providence, R. I., 1967.
14. A. F. Bobrova, *Estimates of accuracy of an asymptotic consolidation of countable Markov chains*, Stability Problems for Stochastic Models (1983), "Vsesoyus. Nauchno-Issled. Inst. Sistem. Issled.", Moscow, 16–24; English transl. in J. Soviet Mat. **32** (1986).
15. N. N. Bogolyubov and S. G. Krein, *On positive quite continuous operators*, Sbirnik Prac Inst. Math. AN USSR (1946), 9–9. (Ukrainian)
16. H. Bonsdorf, *Characterization of uniform recurrence for general Markov chains*, Ann. Acad. Sci. Fenn. ser. A1, Math. Diss. **32** (1980).
17. A. A. Borovkov, *Stochastic processes in the queuing theory*, "Nauka", Moscow, 1972; English transl. Springer–Verlag, Berlin and New York, 1976.
18. ———, *Some bounds of the convergence speed in stability theorems*, Teor. Veroyatnost. i Primenen. **22** (1977), 689–699; English transl. in Theory Probab. Appl. **22** (1977).
19. ———, *Asymptotic methods in the queuing theory*, "Nauka", Moscow, 1980. (Russian)
20. ———, *Ergodicity and stability of the sequence* $w(n+1) = f(w(n), \xi(n))$, Teor. Veroyatnost. i Primenen. **33** (1988), 641–658; English transl. in Theory Probab. Appl. **33** (1988), 595–611.
21. ———, *Ergodicity and stability of multidimensional Markov chains*, Teor. Veroyatnost. i Primenen. **35** (1990), 543–547; English transl. in Theory Probab. Appl. **35** (1990), 542–546.
22. ———, *Lyapunov functions and the ergodicity of multidimensional Markov chains*, Teor. Veroyatnost. i Primenen. **36** (1991a), 93–110; English transl. in Theory Probab. Appl. **36** (1991), 1–18.
23. ———, *Ergodicity conditions for Markov chains that are not connected with irreducibility in the sense of Harris*, Sib. Math. Zh. **32** (1991b), 6–19; English transl. in Siberian Math. J. **32** (1991), 543–554.
24. A. Brunel and D. Revuz, *Quelques applications probabilistes de la quasi-compasite*, Ann. Inst. H. Poincare, ser. B **B10** (1974), 301–337.

25. K. S. Chan, *A note on the geometric ergodicity of a Markov chain*, Adv. in Appl. Probab. **21** (1989), 702–704.
26. M. F. Chen, *Exponential L2-convergence and L2-spectral gap for Markov processes*, Acta Math. Sinica (N.S.) **7** (1991), 19–37.
27. J. E. Cohen, Y. Imasa, G. Rautu, M. B. Ruskai, E. Seneta, and G. Zbaganu, *Relative entropy under mappings by stochastic matrices*, Linear Alg. Appl. **179** (1993), 211–235.
28. K. L. Chung, *Markov chains with stationary transition probabilities*, Springer–Verlag, Berlin, 1960.
29. D. J. Daley, *Upper bounds for the renewal function via Fourier method*, Ann. Probab. **6** (1978), 876–884.
30. ———, *Tight bounds for the renewal function of a random walk*, Ann. Probab. **8** (1980), 615–621.
31. D. J. Daley, A. Y. Kreinin, and C. D. Trengove, *Inequalities concerning the waiting-time in single-server queue: a survey*, Oxford Statist. Sci. Ser. **9** (1992), Oxford Univ. Press, New York, 177–223.
32. Yu. A. Davydov, *Mixing conditions for Markov chains*, Teor. Veroyatnost. i Primenen. **16** (1973), 321–337; English transl. in Theory Probab. Appl. **16** (1973).
33. P. Diaconis and D. Strook, *Geometric bounds for eigenvalues of Markov chains*, Ann. Appl. Probab. **1** (1991), 36–61.
34. W. Doeblin, *Elements d'une theorie generale des chains simples constantes de Markov*, Ann. Sci. Ecole Norm. Super. **57** (1940), 61–111.
35. J. L. Doob, *Stochastic processes*, John Wiley & Sons, New York, 1953.
36. N. Dunford and J. T. Schvartz, *Linear operators, Part I: General theory*, Interscience Publishers, New York, 1953.
37. P. D. Feigin and E. Rubinstein, *Equivalent description of perturbed Markov processes*, Stoch. Proc. Appl. **9** (1979), 261–272.
38. W. Feller, *An introduction to probability theory and its applications*, vol. 2, John Wiley & Sons, New York, 1966.
39. J. A. Fill, *Eigenvalue bounds on convergence to stationarity for nonreversible Markov chains, with an application to the exclusion process*, Ann. Appl. Probab. **1** (1991), 62–87.
40. Y. P. Filonov, *Stability of properties of recurrence and transience of Markov chains*, Teor. Veroyatnost. i Mat. Statist. **19** (1978), 133–143; English transl. in Theory Probab. Math. Statist. **19**.
41. R. E. Fonderic and C. D. Meyer Jr, *Sensitivity of the stationary distribution vector for an ergodic Markov chain*, Linear Algebra Appl. **76** (1986), 1–17.
42. R. Fortet, *Condition de Doeblin et quasi-compasite*, Ann. Inst. H. Poincare, ser. B **B14** (1978), 379–390.
43. P. Franken and D. Stoyan, *Stabilitatsatze fur eine Klasse homogener Markovscher processe*, Math. Nachr. **61** (1974), 311–316.
44. F. R. Gantmakher, *Theory of matrix*, "Nauka", Moscow, 1967. (Russian)
45. I. I. Gikhman and A. V. Skorokhod, *Theory of stochastic processes*, vol. 1–2, "Nauka", Moscow, 1973. (Russian)
46. D. B. Gnedenko and A. V. Solov′ev, *An estimate of reliability of renewed systems*, Izv. Akad. Nauk SSSR Tekhn. Kibernet. **4** (1975), 83–92; English transl. in Engrg. Cybernetics (1975).
47. L. S. Goldenstein, I. C. Gohberg, and A. S. Markus, *Investigation of some properties of linear bounded operators in relation with their q-norm*, Uchen. Zapiski Kishinev. Univers. **29** (1957), 29–36. (Russian)
48. D. Griffeath, *A maximal coupling for Markov chains*, Z. Wahrscheinlichkeits. verw. Geb. **31** (1975), 95–106.
49. P. P. Gudinas, *On improvements of central limit theorem for homogeneous Markov chain*, Litovsk. Math. Sbornik **22** (1982a), 66–68. (Russian)
50. ———, *On B-regularity of a homogeneous Markov chain*, Litovsk. Math. Sbornik **22** (1982), 67–80. (Russian)
51. P. R. Halmos, *Measure theory*, Springer-Verlag, Berlin, 1974.
52. T. E. Harris, *The existence of stationary measures for certain Markov processes*, Proc. 3rd Berkeley Symp. Math. Statist. Probab., Univ. Calif. **2** (1956), 113–124.

53. D. J. Hartfiel, *How a row change in a stochastic matrix affects the corresponding stochastic eigenvector ?*, Linear and Multilinear Algebra **10** (1981), 205–216.
54. ———, *Homogeneous Markov chains with bounded transition matrix*, J. Appl. Probab. **31** (1994), 362–372.
55. D. P. Heyman, *Approximating the stationary distribution of an infinite stochastic matrix*, J. Appl. Probab. **28** (1991), 96–103.
56. E. Hille and R. S. Phillips, *Functional analysis and semigroups*, American Mathematical Society, Providence, 1957.
57. A. Hordijk and F. Spieksma, *On ergodicity and recurrence properties of a Markov chain with an application to an open Jackson network*, Adv. in Appl. Probab. **24** (1992), 343–376.
58. J.J. Hunter, *The computation of stationary distribution of Markov chain through perturbation*, J. Appl. Math. Stoch. Anal. **4** (1991), 29–46.
59. D. Isaacson and G. R. Luecke, *Strongly ergodic Markov chains and rates of convergence using spectral conditions*, Stoch. Proc. Appl. **7** (1978), 113–121.
60. D. Isaacson and R. L. Tweedie, *Criteria for strong ergodicity of Markov chains*, J. Appl. Probab. **15** (1978), 87–95.
61. M. O. Jneid, N. A. Ruzhevich, and V. M. Shurenkov, *Potentials of ergodic chains, semi-Markov processes and sums of nonnegative variables*, Teor. Veroyatnost. i Primenen. **33** (1988), 545–559; English transl. in Theory Probab. Appl. **33** (1988).
62. V. V. Kalashnikov, *An analysis of stability in problems of queuing systems by method of test functions*, Teor. Veroyatnost. i Primenen. **22** (1977), 89–105; English transl. in Theory Probab. Appl. **22** (1977).
63. ———, *Qualitative analysis of behavior of complex systems by method of test functions*, "Nauka", Moscow, 1978. (Russian)
64. ———, *Investigation of continuity of stochastic sequences by construction of " proximity points"*, Stability Problems for Stochastic Models (1980), "Vsesoyus. Nauchno-Issled. Inst. Sistem. Issled.", Moscow, 52–57; English transl. in J. Soviet Mat.
65. ———, *Methods of estimation of continuity in problem of states consolidation*, Stability Problems for Stochastic Models (1981), "Vsesoyus. Nauchno-Issled. Inst. Sistem. Issled.", Moscow, 48–53; English transl. in J.Soviet Mat.
66. ———, *Estimates of continuity in queuing models and related problems*, IV Internat. School on Probab. Theory and Math. Statist., Varna (1982), "Bulg. Ac. Sci.", Sofia, 119–166.
67. ———, *An estimate of the speed of convergence in Renyi's theorem*, Stability Problems for Stochastic Models (1983), "Vsesoyus. Nauchno-Issled. Inst. Sistem. Issled.", Moscow, 48–57; English transl. in J. Soviet Mat. **32** (1986).
68. ———, *Quantitative estimates in reliability theory*, Math. and Cybernetics **10** (1989), "Znanie", Moscow. (Russian)
69. V. V. Kalashnikov and S. T. Rachev, *Mathematical methods of construction of stochastic queuing models*, "Nauka", Moscow, 1988. (Russian)
70. V. V. Kalashnikov and S. Y. Vsekhsvyatski, *On the connection of Renyi's theorem and renewal theory*, Lect. Notes Math. **1412** (1989), Springer–Verlag, Berlin, 83–102.
71. S. Karlin, *A first course in stochastic processes*, Academic Press, New York, 1968.
72. N. V. Kartashov, *Exponential asymptotics of a Markov renewal matrix*, Asymptotic Problems for Random Processes, Prepr. No 77-24 (1977), "Inst. Math. AN Ukrain. SSR", Kiev, 28–43. (Russian)
73. ———, *Strongly stable Markov chains*, Stability Problems for Stochastic Models (1981a), "Vsesoyus. Nauchno-Issled. Inst. Sistem. Issled.", Moscow, 54–59; English transl. in J. Soviet Mat. **34** (1986), 1493–1498.
74. ———, *On strong stable Markov chains*, Teor. Veroyatnost. i Primenen. **26** (1981b), 655–656; English transl. in Theory Probab. Appl. **26** (1981).
75. ———, *A generalization of the Stone representation and necessary conditions of uniform convergence in renewal theorem*, Teor. Veroyatnost. i Mat. Statist. **26** (1982a), 49–62; English transl. in Theory Probab. Math. Statist. **26** (1983), 53–67.
76. ———, *Equivalence of uniform renewal theorems and criteria for their validity*, Teor. Veroyatnost. i Mat. Statist. **27** (1982b), 51–60; English transl. in Theory Probab. Math. Statist. **27** (1983), 55–64.

77. ———, *Criteria for uniform ergodicity and strong stability of Markov chains with a general state space*, Teor. Veroyatnost. i Mat. Statist. **30** (1984), 65–81; English transl. in Theory Probab. Math. Statist. **30** (1985), 71–89.
78. ———, *Asymptotic representations in the ergodic theorem for general Markov chains and their applications*, Teor. Veroyatnost. i Mat. Statist. **32** (1985a), 113–121; English transl. in Theory Probab. Math. Statist. **32** (1986), 131–139.
79. ———, *Bounds of convergence in consolidation theorems of Markov chains*, Internat. Semin. on Probl. Math. Modelir. and Stability Stoch. Models (1985b), Varna, 27–29. (Russian)
80. ———, *Asymptotic expansions and inequalities in stability theorems for general Markov chains under relatively bounded perturbations*, Stability Problems for Stochastic Models (1985c), "Vsesoyus. Nauchno-Issled. Inst. Sistem. Issled.", Moscow, 95–102; English transl. in J. Soviet Mat.
81. ———, *Inequalities in ergodicity and stability theorems of Markov chains on a general state space. I*, Teor. Veroyatnost. i Primenen. **30** (1985d), 230–240; English transl. in Theory Probab. Appl. 30 (1985).
82. ———, *Inequalities in ergodicity and stability theorems of Markov chains on a general state space. II*, Teor. Veroyatnost. i Primenen. **30** (1985e), 478–485; English transl. in Theory Probab. Appl. 30 (1985).
83. ———, *A refinement of ergodicity estimates of renewal processes and Markov chains*, Teor. Veroyatnost. i Mat. Statist. **32** (1985f), 27–33; English transl. in Theory Probab. Math. Statist. **32** (1985).
84. ———, *Inequalities in theorems on consolidation of Markov chains*, Teor. Veroyatnost. i Mat. Statist. **34** (1986), 62–73; English transl. in Theory Probab. Math. Statist. **34** (1987).
85. ———, *Uniform theorems of ergodicity and stability of Markov chains and their applications*, Teor. Veroyatnost. i Primenen. **32** (1987a), 203–204; English transl. in Theory Probab. Appl. **32** (1987).
86. ———, *Estimates of geometric asymptotics of Markov times on homogeneous chains*, Teor. Veroyatnost. i Mat. Statist. **37** (1987b), 66–77; English transl. in Theory Probab. Math. Statist. **37** (1988), 75–88.
87. ———, *Inequalities in Renyi's theorem*, Teor. Veroyatnost. i Mat. Statist. **45** (1991), 27–33; English transl. in Theory Probab. Math. Statist. **45** (1992), 23–28.
88. T. Kato, *Perturbation theory for linear operators*, Springer–Verlag, Berlin, 1966.
89. J. Keilson, *Markov chain models — rarity and exponentiality*, Springer–Verlag, Berlin, 1979.
90. J. G. Kemeny, *Generalization of a fundamental matrix*, Linear Algebra Appl. **38** (1981), 193–206.
91. J. G. Kemeny, J. L. Snell, and A. W. Knepp, *Denumerable Markov chains*, Springer-Verlag, Berlin, 1976.
92. D. P. Kennedy, *The continuity of the single server queue*, J. Appl. Probab. **9** (1972), 370–391.
93. L. Kleinrock, *Queuing systems: Theory*, vol. 1, John Wiley & Sons, New York, 1975.
94. O. V. Komashko, *Generalized potential of a Markov chain*, Teor. Veroyatnost. i Mat. Statist. **43** (1990), 69–72; English transl. in Theory Probab. Math. Statist. **43** (1991).
95. I. P. Kornfeld, Y. G. Sinai, and S. V. Fomin, *Ergodic theory*, "Nauka", Moscow, 1980. (Russian)
96. V. S. Korolyuk, *Stochastic models of systems*, "Nauk. Dumka", Kiev, 1989. (Russian)
97. V. S. Korolyuk and V. M. Shurenkov, *Method of potential in boundary problems for random walks on Markov chain*, Ukrain. Math. Journ. **29** (1977), 464–471. (Russian)
98. V. S. Korolyuk and A. F. Turbin, *Semi-Markov processes and their applications*, "Nauk. Dumka", Kiev, 1976. (Russian)
99. ———, *Mathematical foundations of phase consolidation of complex systems*, "Nauk. Dumka", Kiev, 1978. (Russian)
100. ———, *Markov renewal processes in reliability problems of complex systems*, "Nauk. Dumka", Kiev, 1982. (Russian)
101. I. N. Kovalenko, *An algorithm of asymptotic analysis of a sojourn time of Markov chain in a set of states*, Doklady AN USSR, ser. A **6** (1973), 422–426. (Russian)
102. ———, *Limit theorems of reliability theory*, Cybernetics **6** (1977), 106–116. (Russian)

103. ———, *Analysis of rare events in estimating the effectiveness and reliability of systems*, "Sovetskoe Radio", Moscow, 1980. (Russian)

104. ———, *Rare events in queuing theory — a survey*, Queuing Systems Theory Appl. **16** (1994), 1–49.

105. I. N. Kovalenko and N. Yu. Kuznetsov, *Construction of embedded renewal process for essentially high-dimensional processes of queuing and their applications*, Prepr. Inst. Cybern. AN USSR, No 80-12 (1980), Kiev, 1-61. (Russian)

106. I. N. Kovalenko, N. Yu. Kuznecov, and V. M. Shurenkov, *Stochastic processes, Reference book*, "Nauk. Dumka", Kiev, 1983. (Russian)

107. E. L. Kozarovitsky, *Uniform ergodicity and stability of semi-Markov processes*, Teor. Veroyatnost. i Mat. Statist. **44** (1991), 67–72; English transl. in Theory Probab. Math. Statist. **44** (1992).

108. M. G. Krein and M. A. Rootman, *Linear operators that keep invariant cone in a Banach space*, Uspekhi Math. Nauk **3** (1948), 3-95. (Russian)

109. S. G. Krein, et. al, *Functional analysis*, "Nauka", Moscow, 1972. (Russian)

110. G. M. Laslet, D. B. Pollard, and R. L. Tweedie, *Techniques for establishing ergodic and recurrence properties of continuous valued Markov chains*, Nav. Res. Log. Quart. **25** (1978), 455–472.

111. A. Lesanovsky, *Coefficients of ergodicity generated by nonsymmetrical vector norms*, Czechoslovak Math. J. **110(115)** (1990), 284-294.

112. B. A. Lifshits, *On central limit theorem for sum of random variables connected in a chain*, Teor. Veroyatnost. i Primenen. **23** (1978), 295–312; English transl. in Theory Probab. Appl. **23** (1978).

113. B. Lindquist, *Ergodic Markov chains with finite convergence time*, Stoch. Proc. Appl. **11** (1981), 91-99.

114. H. P. Lotz, *Uniform ergodic theorems for Markov operator on $C(X)$*, Math. Zeitschr. **178** (1981), 145-156.

115. G. Louchard and G. Latouche, *Geometric bounds on iterate approximation for nearly completely decomposable Markov chains*, J. Appl. Probab. **27** (1990), 521–529.

116. V. A. Malyshev and M. V. Menshikov, *Ergodicity, continuity and analyticity of countable Markov chains*, Trudy Moscow. Math. Obszhestva **39** (1979), 3–48. (Russian)

117. Y. I. Mazhuga, *An estimate of killing moment of continuous homogeneous Markov process*, Teor. Veroyatnost. i Mat. Statist. **39** (1988), 83–87; English transl. in Theory Probab. Math. Statist. **39** (1989).

118. ———, *Estimates of sojourn time in subset of states of semi-Markov process*, Kibernetika **2** (1989), 122–123. (Russian)

119. C. D. Meyer, *The condition of a finite Markov chain and perturbation bounds for the limiting distribution*, SIAM J. Algebraic Discrete Methods **1** (1980), 273–287.

120. ———, *The character of a finite Markov chain*, IMA Vol. Math. Appl. **84** (1993), Springer-Verlag, New York, 47–58.

121. S. P. Meyn, *Stability of Markov chains on topological spaces with applications to adaptive control and time series analysis*, Lect. Notes in Control and Inform. Sci. **161** (1991), Springer–Verlag, Berlin, 369–401.

122. S. P. Meyn and R. L. Tweedie, *Stability of Markov processes. I. Criteria for discrete-time chains*, Adv. in Appl. Probab. **24** (1992), 542–574.

123. ———, *Stability of Markov processes. II. Continuous time processes and sampled chains*, Adv. in Appl. Probab. **25** (1993), 487-517.

124. ———, *State-dependent criteria for convergence of Markov chains*, Ann. Appl. Probab. **4** (1994), 149–168.

125. S. V. Nagaev, *Some limit theorems for homogeneous Markov chains*, Teor. Veroyatnost. i Primenen. **2** (1957), 389–416; English transl. Theory Probab. Appl. **2** (1957).

126. ———, *A refinement of limit theorems for homogeneous Markov chains*, Teor. Veroyatnost. i Primenen. **6** (1961a), 67–86; English transl. in Theory Probab. Appl. **6** (1961).

127. ———, *Some questions of theory of homogeneous Markov processes with discrete time*, Doklady AN SSSR **139** (1961b), 34–36. (Russian)

128. ______, *On ergodic theorem for homogeneous Markov chains*, Doklady AN SSSR **263** (1982), 27–30. (Russian)
129. ______, *Ergodic theorems for homogeneous Markov chains*, Dokl. Acad. Nauk SSSR **306** (1989), 283–286; English transl. in Sov. Math. Dokl. **39** (1989), 483–486.
130. M. F. Neuts and J. L. Teugels, *Exponential ergodicity of the M/G/1 queue*, SIAM J. Appl. Math. **17** (1975), 921–929.
131. J. Neveu, *Mathematical foundations of the calculus of probability*, Holden Day, San Francisco, 1964a.
132. ______, *Chaines de Markov et theorie du potentiel*, Ann. Faculte des Science, Clermont **3** (1964b), 37–65.
133. ______, *Potentiel Markovien recurrent des chaines de Harris*, Ann. Inst. Fourier, Grenoble **22** (1972), 85–130.
134. I. M. Novikov and A. F. Turbin, *Potential operators of Markov chains that have positive point of recurrence*, Applicat. of Analytical Methods in Probabilistic. Problems (1986), "Inst. Math. AN USSR", Kiev, 88–99. (Russian)
135. E. Nummelin, *The convergence of sums of transition probabilities of a positive recurrent Markov chain*, Math. Scand. **48** (1981), 79–95.
136. ______, *General irreducible Markov chains and non-negative operators*, Cambridge Univ. Press, London, 1984.
137. E. Nummelin and P. Tuominen, *Geometric ergodicity Harris recurrent Markov chains with applications to renewal theory*, Stoch. Proc. Appl. **12** (1982), 187–202.
138. E. Nummelin and R. L. Tweedie, *Geometric ergodicity for a class of Markov chains*, Ann. Sci. Univ. Clermont, ser. Math **61** (1976), 145–154.
139. ______, *Geometric ergodicity and R-positivity for general Markov chains*, Ann. Probab. **6** (1978), 404–420.
140. C.A. O'Cinneide, *Entry wise perturbation theory and error analysis for Markov chains*, Numer. Math. **65** (1993), 109–120.
141. S. Orey, *Potential kernels for recurrent Markov chains*, J. Math. Anal. Appl. **8** (1964), 104–132.
142. ______, *Limit theorems for Markov chains transition probabilities*, Van Nostrand, New York, 1971.
143. A. Plocki, *Convergence d'un vecteur invariante d'une matrice stochastique* $Q = (1-e)Q_1 + eQ_2$ *ou* $Q_1T = Q_1$, Wyz. Szkola Ped. Krakov. Rochnik Prace, ser.Math **9** (1979), 119–123.
144. Y. D. Plotkin and A. F. Turbin, *Conversion of perturbed on spectrum normal solvable operators*, Ukrain. Math. Journ **27** (1975), 477–486. (Russian)
145. N. N. Popov, *Geometrical ergodicity of Markov chains with general state space*, Doklady AN SSSR **247** (1979), 798–802. (Russian)
146. Yu. V. Prokhorov, Yu. A. Rozanov, *Probability theory, Basic notions, Limit theorems, Stochastic processes.*, "Nauka", Moscow, 1973. (Russian)
147. D. Revuz, *Quasi-compasite de certains operateurs lies aux chaines de Harris*, Asterisque **4** (1973), 127–136.
148. ______, *Markov chains*, North–Holland Elsevier, Amsterdam, 1975.
149. W. Rudin, *Functional analysis*, McGraw Hill, New York, 1973.
150. L. Schwartz, *Analysis mathematique, parts* 1, 2, Hermann, Paris, 1967.
151. E. Seneta, *Coefficients of ergodicity: structure and applications*, Adv. Appl. Probab. **11** (1979), 576–590.
152. ______, *Sensitivity to perturbation of the stationary distribution: some refinements*, Linear Algebra Appl. **108** (1988), 121–126.
153. ______, *Applications of ergodicity coefficients to homogeneous Markov chains*, Contemp. Math. **149** (1993a), Amer. Math. Soc., Prov., RI, 189–199.
154. ______, *Ergodicity for products of infinite stochastic matrices*, J. Theoret. Probab. **6** (1993b), 345–352.
155. ______, *Explicit forms of ergodicity coefficients of stochastic matrices*, Linear Algebra Appl. **191** (1993c), 245–252.
156. ______, *Sensitivity of finite Markov chains under perturbation*, Statist. Probab. Lett. **17** (1993d), 163–168.

157. B. A. Sevast'anov, *Ergodic theorem for Markov processes and their application to telephone systems with refusals*, Teor. Veroyatnost. i Primenen. **1** (1957), 106–116; English transl. in Theory Probab. Appl. **1** (1957).
158. V. M. Shurenkov, *Ergodic theorems and related problems of the theory of stochastic processes*, "Nauk. Dumka", Kiev, 1981. (Russian)
159. ———, *Markov intervention of chance and limit theorems*, Mat. Sb. **126** (1985), 172–193; English transl. in Math. USSR-Sb. **42** (1985).
160. ———, *Ergodic Markov processes*, "Nauka", Moscow, 1986. (Russian)
161. D. S. Sil'vestrov, *Ergodic theorems in series scheme*, Teor. Veroyatnost. i Primenen. **23** (1978), 878–879; English transl. in Theory Probab. Appl. **23** (1978).
162. A. D. Solov'ev, *Asymptotic behavior of the moment of first appearance of rare event in regenerative process*, Izv. Akad. Nauk SSSR Tekhn. Kibernet. **6** (1971), 79–89; English transl. in Engrg. Cybernetics **9** (1971).
163. F. M. Spieksma and R. L. Tweedie, *Strengthening ergodicity to geometric ergodicity for Markov chains*, Commun. Statist. Stochastic Models **10** (1994), 45–74.
164. G. W. Stewart, *On the sensitivity of nearly uncoupled Markov chains*, Probab. Pure Appl. **8** (1991), "Marcel Dekker", New York, 105–119.
165. D. Stoyan, *Qualitative Eigenschaften und Abschatzungen stochastischer Modelle*, Academie-Verlag, Berlin, 1977.
166. ———, *Comparison methods for queues and other stochastic models*, John Wiley & Sons, New York, 1983.
167. ———, *Coefficients of ergodicity with respect to vector norms*, J. Appl. Probab. **20** (1983), 277–287.
168. E. V. Sugakova, *Exponential asymptotics of sum of geometrical number of independent random variables*, Teor. Veroyatnost. i Mat. Statist. **42** (1990), 135–139; English transl. in Theory Probab. Math. Statist. **42** (1992).
169. ———, *Estimation of the convergence rate for a thinning flow to the Poisson process*, Kibernetika **1** (1991), 128–129. (Russian)
170. ———, *Estimation with weight for the characterization of asymptotics of sums of independent random variables*, Teor. Veroyatnost. i Mat. Statist. **48** (1993), 185–190; English transl. in Theory Probab. Math. Statist. **48** (1994).
171. R. Syski, Studies in Prob., Optim. and Stat. **1** (1992), IOS Press, Amsterdam.
172. C. P. Tan, *A functional form of a particular coefficients of ergodicity*, J. Appl. Probab. **19** (1982), 858–863.
173. E. C. Titchmarsh, *Theory of functions*, Oxford Univ. Press, London, 1952.
174. P. Tuominen and R. L. Tweedie, *Markov chains with continuous components*, Proc. Lond. Math. Soc. **38** (1979), 89–114.
175. A. F. Turbin, *Abstract potential operators of ergodic Markov processes*, Teor. Sluch. Process. **9** (1981), 101–109. (Russian)
176. A. F. Turbin, A. N. Karlat, V. N. Kuznetsov, and M. M. Novikov, *Semi-Markov repairable systems and queuing systems*, "Stiintsa", Kishinev, 1991. (Russian)
177. R. L. Tweedie, *Truncation procedures for non-negative matrices*, J. Appl. Probab. **8** (1971), 311–320.
178. ———, *The robustness of positive recurrence of Markov chains under perturbations of the transition probabilities*, J. Appl. Probab. **12** (1975), 744–753.
179. ———, *Truncation approximation of the limit probabilities for denumerable semi-Markov processes*, J. Appl. Probab. **12** (1975), 161–163.
180. ———, *Perturbations of countable Markov chains and processes*, Ann. Inst. Statist. Math. **32** (1980), 283–290.
181. ———, *The existence of moments for stationary Markov chains*, J. Appl. Probab. **20** (1983), 191–196.
182. Werner Wolf, *Approximation des invarianten Wahrscheinlichkeitsmasses eine Markoffschen Kerns bezuglich Schwacher Konvergenz*, Z. Angew. Math. Mech. **60** (1980), 372–374.
183. V. M. Yadrenko, *An extremal problem for a Markov chain that describes the simplest random walk with a reflection*, Selected Problems of the Recurrent Theory of Random Processes (1988), Inst. Math Akad. Nauk USSR, Kiev, 116–119. (Russian)

184. K. Yosida and S. Kakutani, *Operator-theoretical treatment of Markov's processes and mean ergodic theorem*, Ann. Math. **42** (1941), 188–228.

185. A. I. Zeifman, *Lect. Notes Math.*, vol. 1155, Springer–Verlag, Berlin, 1985, pp. 401–414.

186. ———, *Uniform quasi-ergodicity and stability of inhomogeneous continuous-time Markov chains*, J. Soviet Math. **59** (1992), 911–913.

187. V. M. Zolotarev, *Quantitative bounds in problems of continuity of queuing systems*, Teor. Veroyatnost. i Primenen. **20** (1975), 215–218; English transl. in Theory Probab. Appl. **20** (1975).

188. ———, *Metric distances in spaces of random variables and their distributions*, Mat. Sb. **101** (1976a), 416–454; English transl. in Math. USSR–Sb. (1977).

189. ———, *On stochastic continuity of queuing systems G/G/1*, Teor. Veroyatnost. i Primenen. **21** (1976b), 260–279; English transl. in Theory Probab. Appl. 21 (1976).

190. ———, *Modern theory of summation of independent random variables*, "Nauka", Moscow, 1986. (Russian)

191. A. M. Zubkov, *Inequalities for transition probabilities with taboos and their applications*, Mat. Sb. **109** (1979), 491–532; English transl. in Math. USSR Sb. **37** (1980).

192. ———, *Explicit estimate in the law of large numbers for sequences with mixing*, Doklady AN SSSR **4** (1981), 491–532. (Russian)

Bei Fragen zur Produktsicherheit wenden Sie sich bitte an:
If you have any questions regarding product safety,
please contact:

Walter de Gruyter GmbH
Genthiner Straße 13
10785 Berlin
productsafety@degruyterbrill.com